CROWOOD METALWORKING GUIDES

MILLING

Do not wear loose clothing when operating this machine
SENIOR
TOM
SENIOR
LIVERSEDGE LTD

CROWOOD METALWORKING GUIDES

MILLING

DAVID A. CLARK

First published in 2014 by
The Crowood Press Ltd
Ramsbury, Marlborough
Wiltshire SN8 2HR

enquiries@crowood.com

www.crowood.com

Paperback edition 2023

British Library Cataloguing-in-Publication Data
A catalogue record for this book is available from the British Library.

ISBN 978 0 7198 4313 6

Frontispiece: A Tom Senior light vertical milling machine suitable for use in a small workshop.

Acknowledgements
I would like to thank the following for the use of some of their illustrations: Arc Euro Trade, MyHobbyStore Ltd (*Publishers of Model Engineer* and *Model Engineers' Workshop*), Warco and Ken Willson.

Cover design by Maggie Mellett

Typeset by Servis Filmsetting Ltd, Stockport, Cheshire

Printed and bound in Great Britain by CPI Group (UK) Ltd, Croydon CR0 4YY

Contents

Introduction

This book will assist the reader in developing their milling skills and applying them to make finished pieces of work accurately and to very high standards. If you are an amateur or professional engineer who is perhaps looking to install your first milling machine in your own workshop, this book is for you. It is a complete comprehensive guide to buying, installing and using a small milling machine, with the accessories that you will need to use the milling machine to its full capabilities. The book will teach you milling from the basic principles right through to advanced milling methods. Whether you are working in a home or light industrial environment, you will learn the important milling skills and insider tips from someone who has spent a large part of his life milling in industry.

There are several types of milling machine such as CNC mills, horizontal mills, vertical mills and machining centres. The milling machine we will concentrate on is the vertical mill and the vertical turret mill, which are the ones that will be most likely to be found in the small workshop. Many of the principles of milling apply to the entire range of milling machines so this book will be useful to, say, the horizontal milling machine owner as well as the CNC milling machine owner. The milling machine is very versatile – at a pinch it can also be used as a simple lathe by holding the workpiece in the milling machine spindle and a lathe tool in the machine vice.

ABOUT THE AUTHOR

I spent more than thirty years in the engineering industry, mostly doing milling in all its forms. I was trained to make industrial sewing machine components which, as you can imagine, are mostly very small and were made to very tight tolerances. I also worked on the huge wing holding fixture used by Airbus to transport wings in their Supper Guppy aircraft. I have also trained many people in milling machine practice and engineering in general.

I was the editor of *Model Engineer* magazine for about five years and I edited *Model Engineers' Workshop* from 2007 to 2014; these are the two main UK magazines for machining in the home workshop. This editing experience has given me a unique insight into what amateur engineers need to help them succeed. This book will enable you to get started with a milling machine in your workshop easily and, most importantly, safely.

Model Engineers' Workshop magazine, first published in 1990, is a recent addition to the model engineering scene. It often covers projects and techniques suitable for the milling machine. A new issue goes on sale every four weeks.

The Model Engineer magazine is probably the world's longest running model engineering hobby magazine and is published every fortnight. It is mainly about making precision models but often includes articles about milling machines. The milling machine depicted on the cover of the issue shown is a model of a horizontal milling machine made by Ray Griffin.

1 The Milling Machine

The machines we will be discussing in this book are the vertical milling machine and the vertical turret milling machine. The principles of milling are also applicable to the horizontal milling machine, and some horizontal milling machines can be fitted with a vertical head (when they are known as universal mills). Usually these are fixed heads and do not have a moving quill, however, so the only way to adjust the depth of cut is to raise and lower the mill's table. Also available are slotting heads which bolt on in place of the vertical head, and the tool moves up and down rather than rotating.

The vertical milling machine is basically the same as a turret milling machine, but the turret milling machine is much more versatile. While both types of milling machine have their spindles in the vertical plane, the turret milling machine spindle can usually be tilted to the left or right and sometimes forwards and backwards. The plain vertical milling machine can usually only be tipped to the left and the right, often by tilting the whole vertical column that the head is mounted on.

A typical example of a turret milling machine is the Bridgeport style mill. This is quite a large machine, but many are finding their way into the home workshop. These mills are a standard item in many industrial workshops and are often available at a reasonable price. They do however require a specialist machinery transport company to move them as they are very heavy and are not really suitable for the amateur engineer to move by themselves.

Fig. 1.1 A typical bench mounting vertical milling machine as used in home workshops is the Warco WM 14 milling machine shown here. You can buy purpose-designed cabinet stands for most of the Warco range.

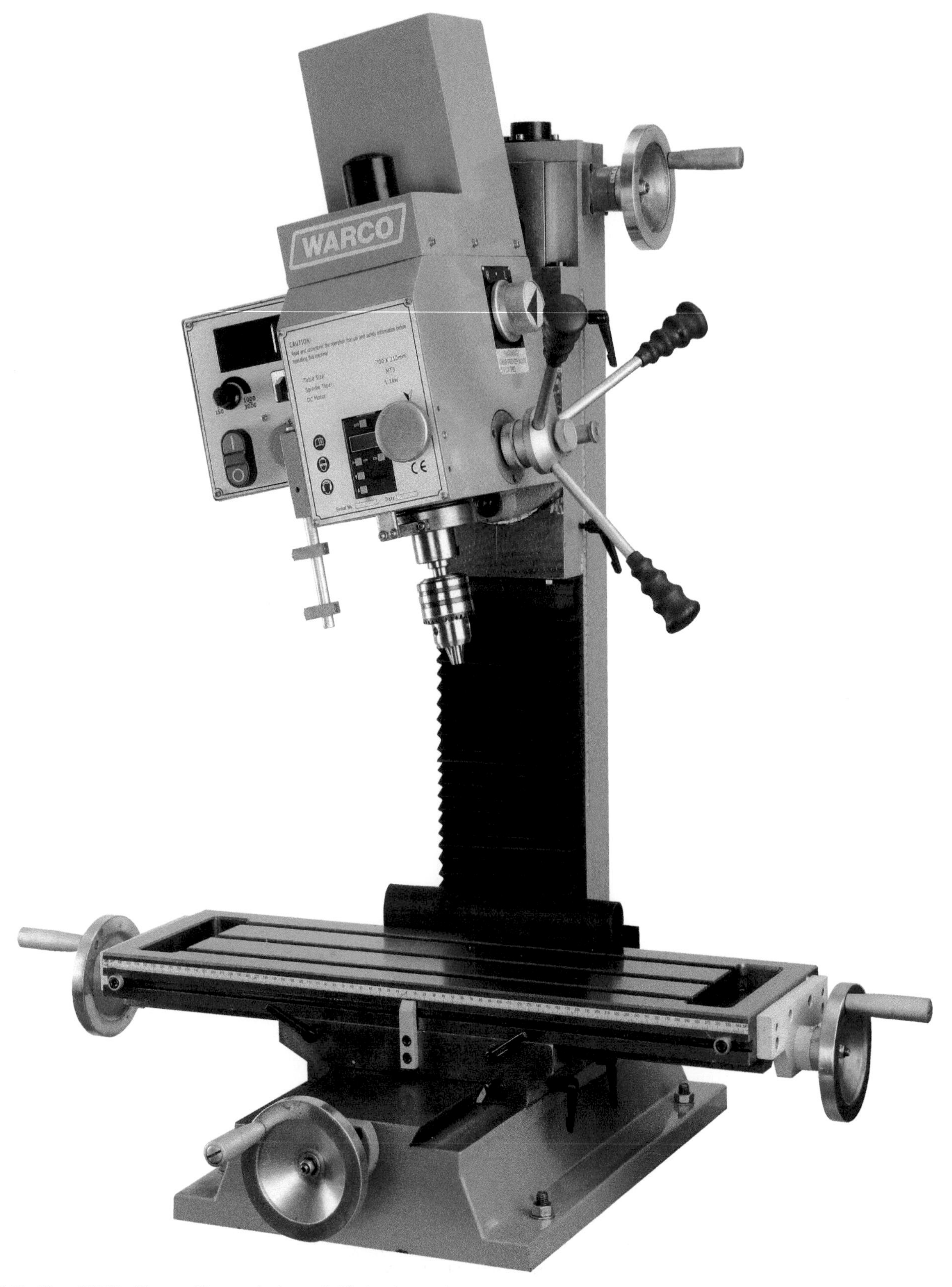

Fig. 1.2 The Warco WM 16 milling machine can also be supplied for bench mounting.

Fig. 1.3 A Bridgeport milling machine can often be found in commercial factories, but more and more are finding their way into home workshops. They usually have a self-contained X axis power feed gearbox for moving the table under power feed.

A typical vertical milling machine will be much smaller than the Bridgeport milling machine and can often be moved by two people, possibly assisted by an engine crane if it is a larger mill. It is usually possible to split the milling machine into two parts: the base unit complete with the machine's table, and the column with the head and motor assembly combined. When you have lifted the base onto the bench or machine stand, you can then add the column and head.

Also available are much smaller turret mills than the Bridgeport, often called VMCs (vertical machining centres). On a turret milling machine the head can usually be twisted to the right and left and rotated around the column, but it cannot usually be tilted forwards and backwards.

Fig. 1.4. This Warco WM 16 milling machine, seen here on its purpose-made stand, is often found in the home workshop.

CHOOSING A MILL

Size

The first thing to decide is: what is the largest piece of work you will be milling on the machine, and what is the largest diameter you will want to hold in the dividing head.

It may be that the longest and widest item you will want to hold is a pair of locomotive mainframes. Perhaps the largest diameter is a traction engine rear wheel for drilling for the spokes and strakes and so on. The locomotive frames can be moved along the table so the table length and movement is not too critical. They can also be rotated 180 degrees to access the far side of the frame plates.

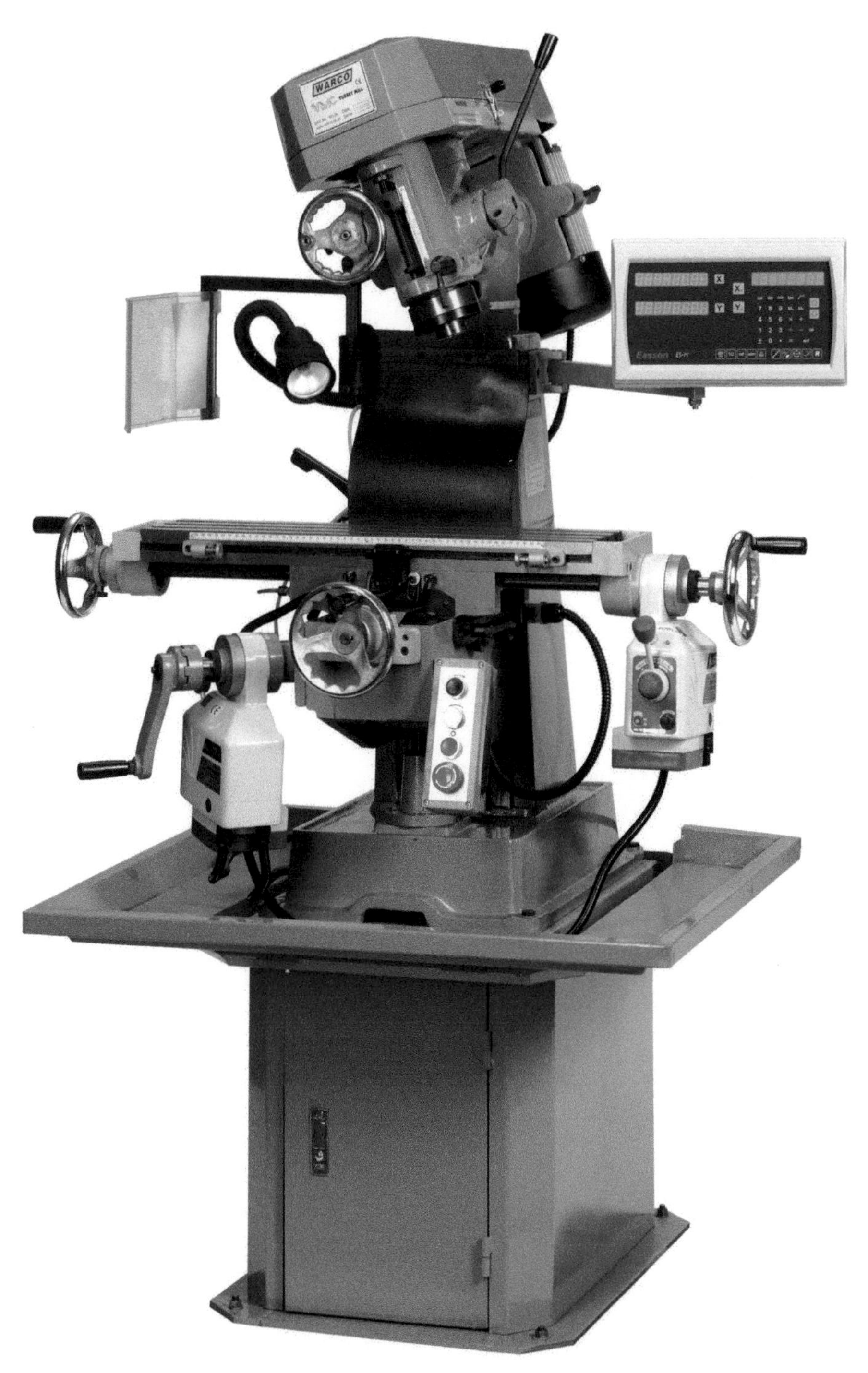

Fig. 1.5. A small turret milling machine will be found to be very useful in the small workshop. This example by Warco is typical of this type of machine. It is quite easy to fit a digital readout, a power feed attachment and a machine light to improve on the machine's facilities.

A Bridgeport milling machine will usually have an X axis travel of 26in (660mm) and a Y axis travel of 12in (300mm) although I think this may vary slightly on some models. This will be sufficient for all but the largest models.

A typical medium-size, hobby-type milling machine will have an X axis travel of about 19in (485mm) and a Y axis travel of 7in (175mm). A small milling machine will have an X axis travel of about 13in (330mm) and a Y axis travel of 5.5in (145mm). These dimensions will vary with different manufacturers and should be considered a starting point only.

When mounted in the dividing head, the traction engine wheel can be almost twice the diameter of the Y axis movement, as the wheel rim can be rotated so that every part of the wheel can be beneath the centre of the milling machine spindle.

You will also be able to do larger work on a turret milling machine than on the vertical milling machine, as often the milling head can be moved in and out on the machine's ram. (The ram is a large dovetailed slide on the top of the milling machine's column that allows the milling machine's head to move in and out and often to rotate.) In this case, you leave the work in the same position on the table and re-datum the work after moving the head on the ram. Often, the ram can pivot to the left and to the right to access all areas of the workpiece. If you are machining a large traction engine wheel rim on a turret milling machine where the head can swivel to the side, you can place the wheel on the table so that the table is in the centre and completely surrounded by the wheel.

Speed

The speeds required for milling will depend on the work you are doing, but unless you are doing very fine work such as engraving, a maximum speed of about 2,000rpm will be more than sufficient. A lot of the modern small mills will have an electronic speed control, sometimes combined with a high/low gearbox. A Bridgeport milling machine will often have a two-speed motor as well as back gear to reduce the speed. Another type of Bridgeport head has a variable speed which is altered by turning a hand wheel.

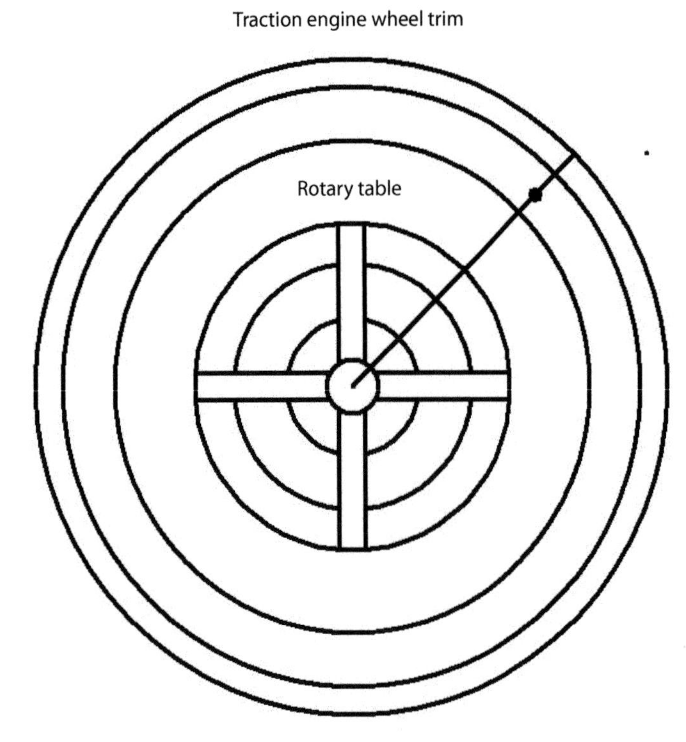

Fig. 1.6. You can use a rotary table to drill a traction engine wheel for the spoke holes. Depending on how you position the rotary table and the milling table you can drill holes at a large radius on a modest size mill. This illustration shows the spoke hole set at 45 degrees from the centre of the table. The machine's spindle is static; only the rotary table and the wheel rim will rotate to drill further holes. You will have to use some extension bars to clamp the traction engine rim to the rotary table.

Fig. 1.7. You can use a dividing head to drill a traction engine wheel for the spoke holes. If you tilt the milling machine head at 45 degrees and the dividing head at 45 degrees you can mill and drill quite large diameters even on a smallish mill.

MILLING MACHINE ELECTRICS

With single-phase electrics you plug the mains lead from the milling machine straight into the mains supply in your house or workshop. The motor is fed from the live, neutral and earth leads. Mills built for industrial use almost always have three-phase motors. Three-phase motors need three live wires. Each wire will be on a different phase. Each phase is like an incoming wave on the seashore, with three waves, one after the other, making up the three phases required. Each phase is the same voltage and current as the preceding one, but each phase takes up a third of each cycle so three sequential phases add up to each complete cycle.

A Bridgeport turret milling machine will almost certainly have three-phase electrics and, unless you are lucky enough to have a three-phase supply to your workshop, it will require a single to three phase converter.

Some three-phase motors will run from a single-phase supply via a converter or an inverter. To install a converter or inverter the converter/inverter maker's instructions should be followed.

Inverters

Some change-over tags may be fitted inside a motor to change it from 440V three-phase down to 240V three-phase. In this case, the motor information plate will often say 240V/440V. If the motor plate says this, the motor is almost certainly capable of running on an inverter; it is just a matter of changing over a few connections inside the motor. This is usually an easy job involving moving some brass or copper links from one terminal to another. Instructions are usually on, or inside, the motor's cover. The three-phase inverter or converter setup will give a quieter and smoother running motor compared to a motor running on a single-phase supply. The inverter will need a little bit of programming to set it up, but if you follow the manufacturer's instructions it will only be a five-minute job to get the motor and inverter combination working properly.

Although the inverter runs at a constant voltage the frequency can be varied. This variation changes the motor speed as the inverter frequency is raised or lowered. I would limit the inverter speed variation from half speed up to full speed so that the motor is still cooled by the fan and does not overheat. You can change the main speed steps using the belts as you normally would. (If you wanted more speed variation, you could fit a separate computer cooling fan to give increased airflow through the motor to keep the temperature of the motor within its limit.)

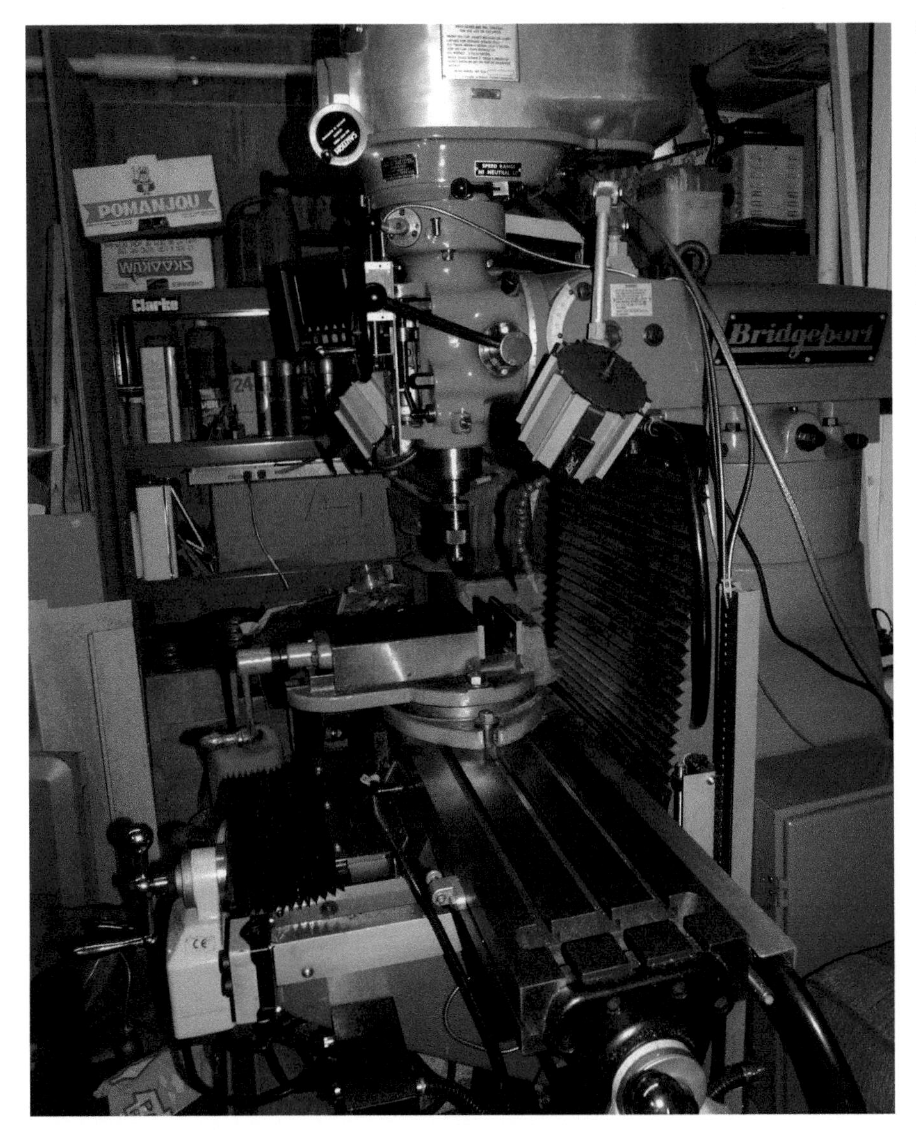

Fig. 1.8. On a Bridgeport mill, the ram can be moved right in to work on one side of the work.

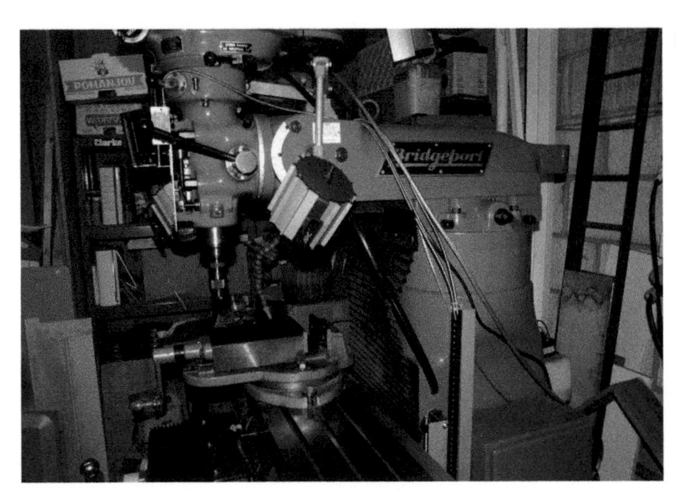

Fig. 1.9. The ram can also be moved out to work on larger workpieces.

Converters

The two main types of converter are the rotary converter and the static converter. The rotary converter will run more than one motor at a time, while the static converter is only designed to run one motor at a time. If you want to run two or more machines at the same time or perhaps run two motors on one machine, perhaps the main machine motor and a motor for the machine's coolant pump or power feed system, then go for the rotary converter.

Rotary converters cost three to four times more than an inverter depending on the size and power output required. If you are running one machine at a time, buy the slightly cheaper static converter. If you fit the static converter with a three-phase socket connector, you can plug in your different machines one at a time. This will suit many home workshop users because for safety reasons you should only run one machine at a time.

SAFETY REGULATIONS

There are several manufacturers in the UK who supply converters and inverters. Make sure you get one that is CE marked and complies to the EMC regulations, the low voltage directive and BS EN ISO 61000-3-2: 2006.

To reverse a three-phase machine if it is running in the wrong direction, just change over any two of the phases by reversing any two of the power wires.

CONVERTER RATING

Rotary and static converters must not exceed their maximum rating. If you are not sure of the load capacity limits required to power your machine(s) the converter manufacturer or supplier should be able to help you.

Fig. 1.10. You can also swivel the ram round to the left or right to get at large workpieces.

Fig. 1.11. If you have a particularly large rim to drill or machine, you can swivel the machine's head and ram away from the knee of the machine and put the rim over the surrounding table. You will have to use a location fixture, possibly locating from the previously drilled spoke holes, to drill the strake holes. It takes a bit of figuring out but it can be done. I had a rim about 2ft diameter for a packaging machine which required about 42 tapered slots milled in it, and after a lot of thought this was the method I came up with to do it. It worked perfectly well.

OTHER FEATURES

A digital readout is often seen as an extra add-on to a milling machine, but I consider it an essential feature. Whether you use a simple scale system similar to a digital vernier scale or a full readout with glass or magnetic scales, you should always buy a machine with a readout fitted or fit a digital readout to your mill. You will never regret it (simple instructions for fitting a digital readout to a milling machine are given in Chapter 2). Although the readout shown was fitted to a specific machine, the principles of fitting the readout apply to many different milling machines.

A power feed and a soluble cutting oil supply are desirable, but by no means essential, features on any milling machine.

THE PARTS OF A MILL

The drive motor

As mentioned previously, an industrial machine's main motor can often work from single or three-phase electrics. However, the majority of milling machines sold specifically for home workshop use will have single-phase motors.

The spindle and head

The motor usually drives the spindle via a single drive belt with multi-step pulleys on the motor and spindle. Sometimes there is a gearbox between the motor and spindle containing the back gear, which allows for a large reduction in spindle speeds. Other

Fig. 1.12. This Bridgeport milling machine head has a variable speed, controlled by the handwheel on the middle left. The air-powered tool at the top of the machine's head is a home-made power draw bar to tighten and remove collets and chucks, etc.

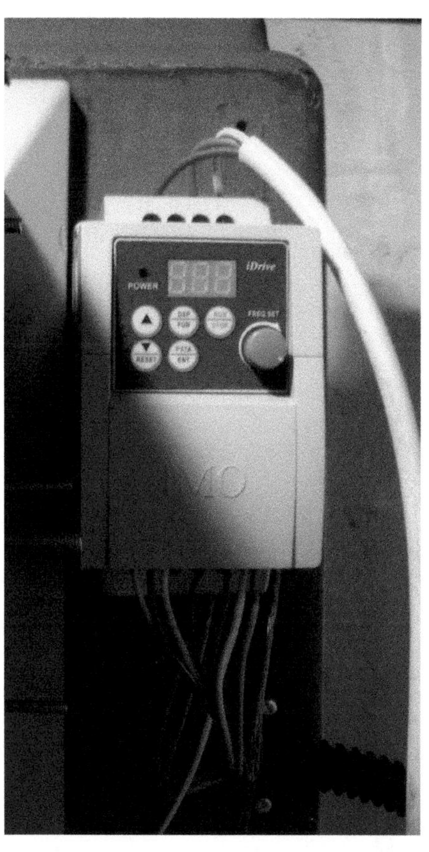

Fig. 1.13. This inverter converts 240 volts single-phase to 240 volts three-phase.

Fig. 1.14. You will find a forward/reverse control box with a built-in speed controller is essential to control the motor.

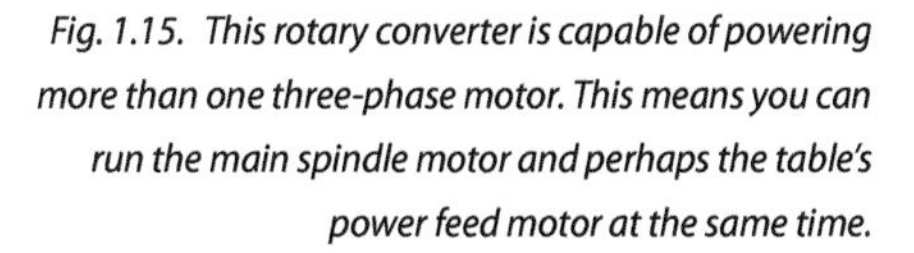

Fig. 1.15. This rotary converter is capable of powering more than one three-phase motor. This means you can run the main spindle motor and perhaps the table's power feed motor at the same time.

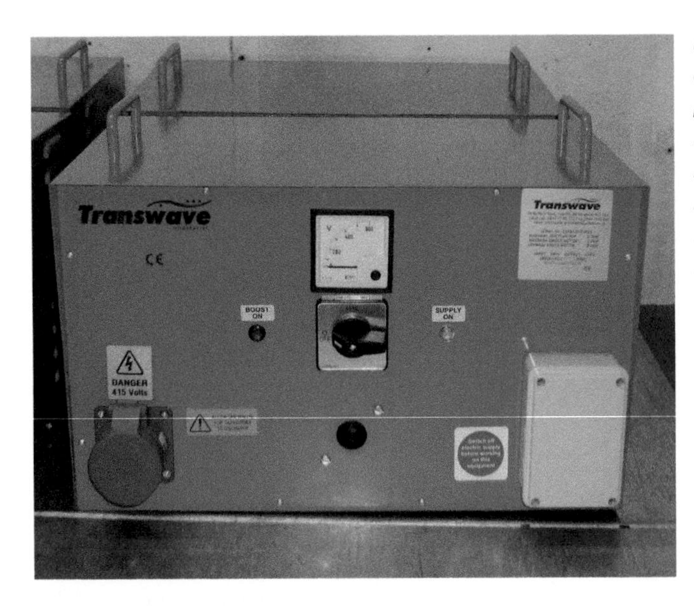

Fig. 1.16. This static converter is suitable for powering one three-phase motor. It has a power socket on the bottom left so that different individual machines can be plugged into it (one at a time).

Fig. 1.20. This is a three-phase motor. The outwards appearance is no different to a single-phase motor other than that many single-phase motors will have a capacitor to make the motor run in the correct direction.

Fig. 1.17. Soluble oil pumps and reservoirs are readily obtainable. Make sure the pump you buy is suitable for the coolant you will be using.

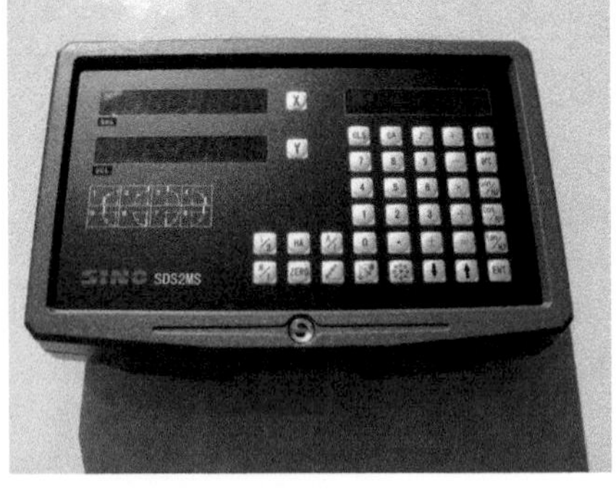

Fig. 1.18. A digital readout is essential to milling work. If at all possible buy a milling machine ready fitted with a readout or at least fit your own readout, it is not a difficult job to fit it and you will not regret it.

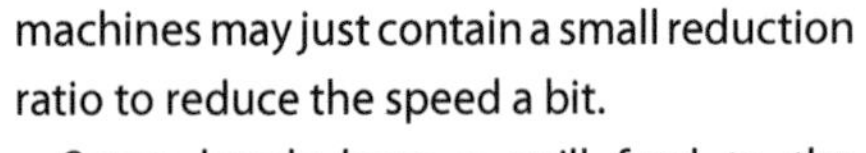

machines may just contain a small reduction ratio to reduce the speed a bit.

Some heads have a quill feed to the spindle so the spindle depth can be wound up and down to set the Z height. Larger mills often have a power down-feed to the quill for use when drilling, reaming and boring. These down-feeds usually have a knock-off system so the down-feed stops automatically when it reaches the stop.

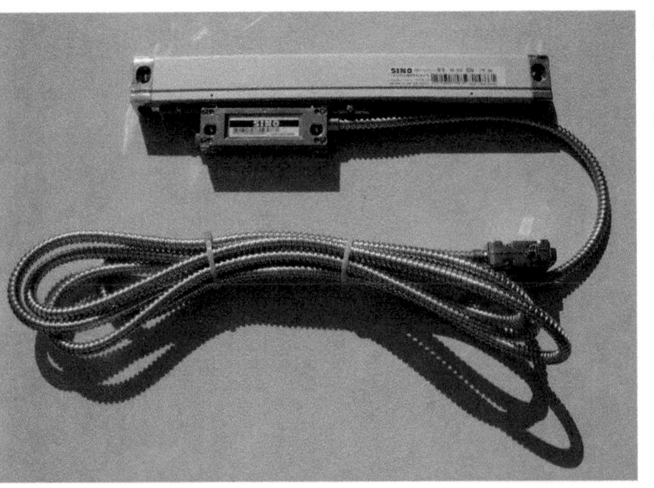

Fig. 1.19. Glass readout scales are available in many different lengths. Ask the readout supplier what length will be suitable for your milling machine.

The column

The milling machine's column is usually a substantial iron casting, although ground steel tubes have often been used as columns. There are two main types of milling machine column: the round column and the dovetailed column. The dovetailed column is usually quite substantial, and the head does not move out of alignment when it is raised or lowered. The round column however does not usually retain alignment when the head is raised or lowered, and this means that you will have to reset the

Fig. 1.21. This is a typical multi-step pulley reduction belt drive, as fitted to many milling machines. You move the belt from step to step to change the mill's speed.

Fig. 1.23. This milling machine has a dovetail column.

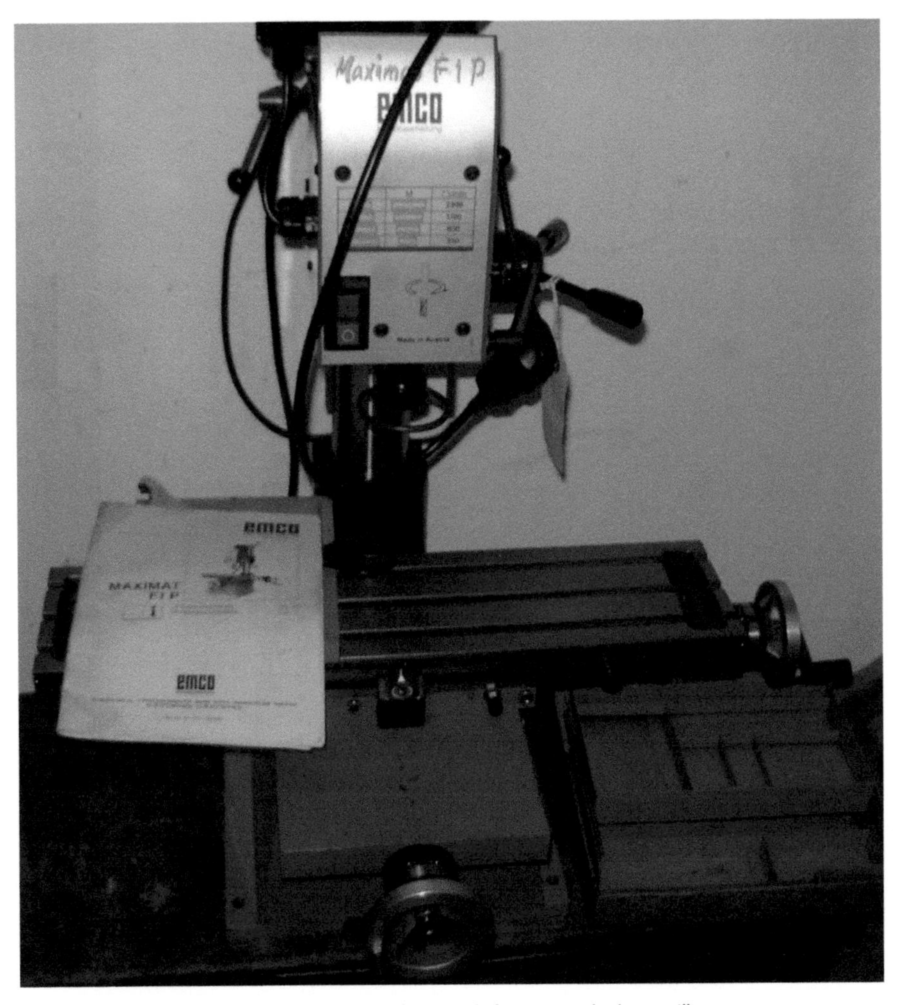

Fig. 1.22. This Emco Maximat F1P milling machine is typical of many round column mills.

X and Y datums when you raise or lower the head on a round column machine. The easiest way to do this is with a wobbler and a wobbler extension. When you have the milling machine head raised high, perhaps because you are drilling with a long drill, you can use the wobbler in an extended adaptor, and when you lower the milling machine head you can re-wobble the datum using a standard wobbler without the extension.

The column of a dovetailed machine can sometimes be set on an angle for angular milling, but a round column that can tilt is very rare.

The knee

Not every milling machine has a knee. Most turret mills will have one. The knee is a large casting that slides up and down the dovetailed column. The machine table is mounted on top of the knee. The knee can be wound up and down independently of the milling machine's head to set the Z depth. If you wish, you can fit a digital readout to the knee as well as to the spindle down-feed to measure the Z travel. Power feeds are also available to fit some milling machine knees. Power feeds certainly ease the back-breaking winding up and down of the heavy knee and can also be used for drilling and boring.

Fig. 1.24. This large casting is called the knee. It allows the milling table to be raised and lowered. The left-hand wheel raises and lowers the knee while the right-hand handle winds the cross slide in and out.

The table

The milling machine table normally moves in the X and Y axes; the table usually travels a far greater distance in the X direction than it does in the Y direction. Some universal milling machine tables swivel left or right in the middle, but you are unlikely to come across this feature; it was often used on universal horizontal and vertical mills where the table leadscrew was connected to a compound dividing head through a gear train for milling helical cutters and gears.

The mill's base

A vertical milling machine will usually have a separate base, which the table is fitted to and the column bolts onto or fits into. The base of a round column milling machine will often allow the column to raise and lower right through the base casting and also through the workbench or stand to change the height of the head.

The base of a turret milling machine is much more likely to have been cast all in one with the column, and it will be a much more substantial item. Quite often the base of the turret milling machine will be used as a soluble oil tank. Sometimes the milling machine base will be a fabricated steel structure.

SO WHAT SHOULD I BUY?

If you have the room, a turret milling machine such as a Bridgeport or the smaller Warco or Chester 626 type milling machine will be ideal. If you can only afford, or indeed only need, a small milling machine, a dove-tailed column milling machine will be the way to go. If finances are tight, a second-hand round column milling machine will do the job but may involve more setting up.

THE A TO Z OF A MILLING MACHINE

- ***A. The machine's two speed motor.***
- ***B. Forward reverse and two speed switch.***
- ***C. Variable speed control handwheel.***
- ***D. Belt drive housing.***
- ***E. Spindle down feed lever control.***
- ***F. Digital readout.***
- ***G. Low voltage machine light.***
- ***H. Spindle down feed control.***
- ***I. The spindle down stop.***
- ***J. Table handwheels.***
- ***K. Left-hand table feed stop.***
- ***L. Electrical switch for feed knockoff.***
- ***M. Machine table.***
- ***N. Right-hand table feed stop.***
- ***O. Feed units for table, cross slide and vertical feed.***
- ***R. Main control panel for all of the electrics.***

Fig. 1.25. From top to bottom, the A to Z of a milling machine.

2 Buying the Mill

BUYING NEW

Suppliers

There are several suppliers of milling machines that advertise regularly in the UK and many more suppliers in various countries around the world. Many of these mills all come from the same Chinese factories and just have slightly different features and paint jobs. A list of suppliers is included at the end of this book.

You should look carefully at the company you are considering purchasing a milling machine from. The cheapest milling machine may not necessarily be the best but neither may the most expensive one either. Look at the price by all means but also look at the after sales service, the availability of accessories and whether they can supply replacement spindle gears and/or speed control boards. A machine with a broken gear or a duff speed controller won't do a lot of milling.

There are several model engineering exhibitions held in the UK every year where you can go and view the various makes of milling machine and compare their size and features. No doubt there are many exhibitions of this type throughout the world where you can get a 'hands on' experience with a milling machine. Do be aware that many of these suppliers will run out of stock during these exhibitions and it could be a two to three month delivery before new stock arrives from China. If you have decided on the milling machine you want, do try and order it prior to the next

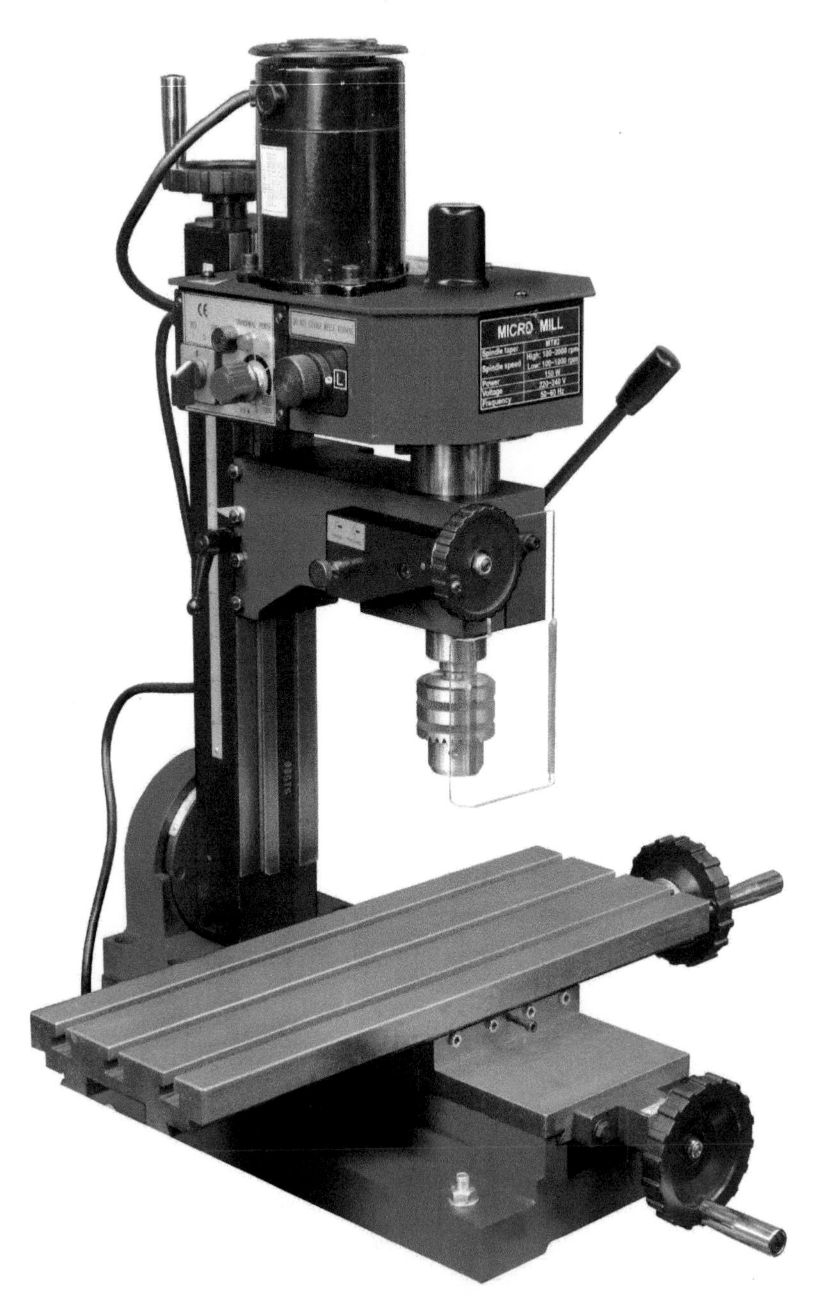

Fig. 2.1. This Sieg Super X1L milling machine (long table) has a geared drive and brushed motor. The long table version is only available from Arc Euro Trade in Leicester (see Useful Contacts at the end of the book).

Fig. 2.2. The Sieg Super X1LP HiTorque milling machine is driven by belt drive from a brushless motor. It also has a longer table as standard.

Fig. 2.3. Sieg Super X2P HiTorque belt drive milling machine with brushless motor and extra-long table.

model engineering exhibition to ensure availability.

Questions to consider

Look on the internet forums and ask questions. The largest UK forum for model engineers and home workshop enthusiasts is *www.model-engineer.co.uk*. This forum has a worldwide following and most questions will be answered by a practising model engineer within a few hours. The membership of this site has many experts amongst them including many professional engineers with up-to-date knowledge of industrial practice and new methods of machining techniques.

Find out who sells the milling machine you are thinking of buying and what, if any, are the limitations. Phone up the supplier and ask for the after sales and service department. Find out how long it takes them to answer the phone and ask them a simple question; did they know the answer or did they have to find out and call you back? Did they even bother to call you back?

Is the machine imperial or metric? Are the leadscrews round figures; that is, if you turn the handle one turn, will it move a specific amount, say 2mm, or is it an odd figure like 1.43mm because the leadscrew is imperial on a so-called metric machine? Ideally you want the machine to move a standard figure per handwheel revolution rather than an odd one. The dial should be engraved to suit the leadscrew movement, not an arbitrary number of divisions that bears no relationship to the leadscrew pitch.

Compare the total price, like for like, from different suppliers. Add up the basic cost of the machine, any accessories you need such as a vice, the carriage price and make sure the VAT is included in the total. You can of

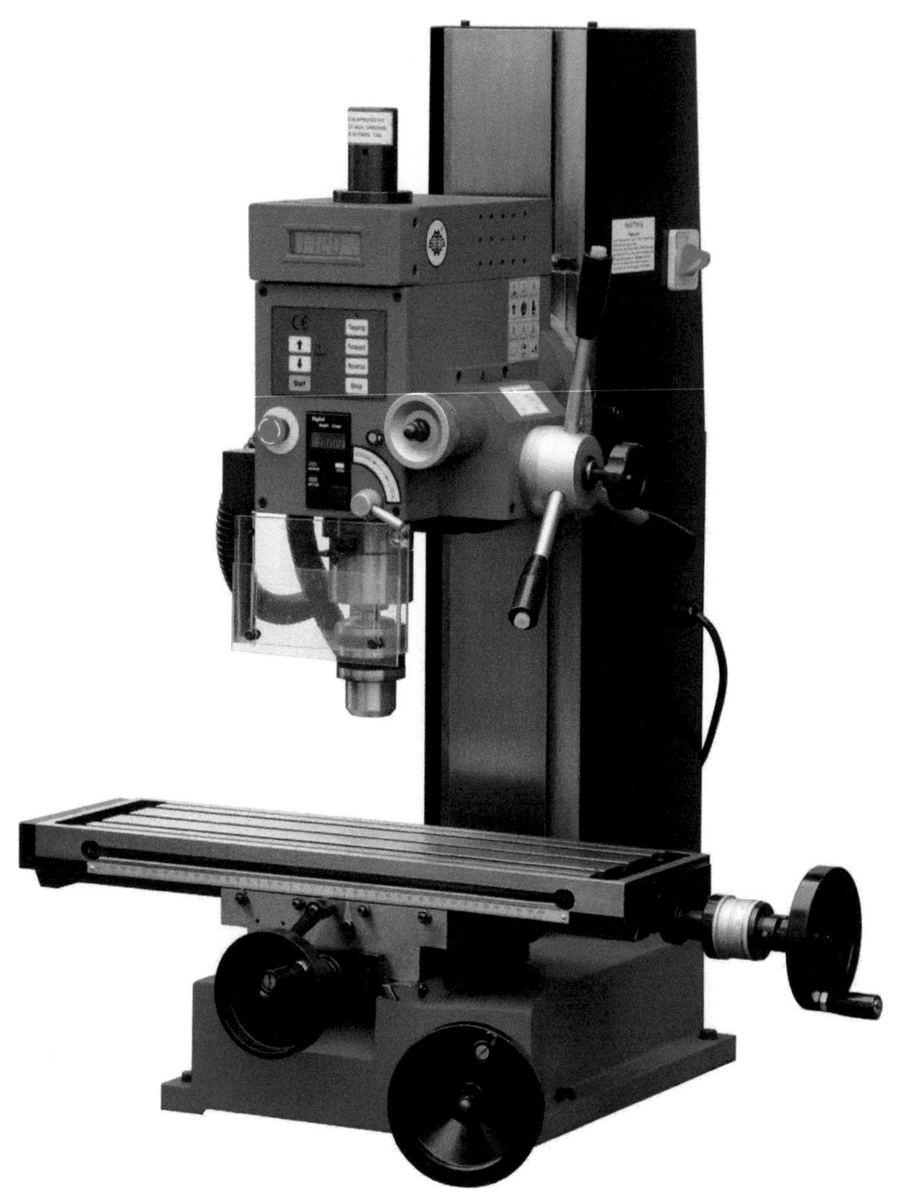

Fig. 2.4. Sieg Super X3 HiTorque direct belt drive milling machine with brushless motor. This machine has a swivelling head and a digital readout to the Z axis, for ease and versatility in use

course buy the machine from one supplier and the tooling from another supplier or even several other suppliers.

Remember to budget for accessories such as a collet chuck and collets to hold milling cutters, a drill chuck to take drills and reamers and, if you need them, the cost of a rotary table and a dividing head. Yes, you will probably need one of each. Cutters are not cheap but, with care, they can last a long time before going blunt. Cutters can be sharpened but they would need to be sent out to a specialist tool and cutter grinder unless you buy or make your own cutter grinder and are prepared to invest some time in learning how to sharpen them. You can buy a lot of new cutters on Ebay for the cost of a tool and cutter grinder.

BUYING SECOND-HAND

The main problem buying second-hand will be the availability of spares, especially for the older milling machines. On the other hand, milling machines do not tend to wear out as fast as lathes in the home workshop environment, especially the British-made ones that are often available on the second-hand market.

MY MILLING MACHINES

The milling machine that can be seen in many of the illustrations in this book is a Tom Senior light vertical milling machine. I bought it second-hand when I had just sold my Chinese milling machine. There was nothing wrong with the Chinese mill – I just did not want to strip it down and move it if I moved house, as it was so heavy. When I originally purchased it, it was delivered to the pavement outside my house and I had to strip it down by myself and move it into a shed in my garden; as a result I did not get round to reassembling it for more than a year.

A couple of years later, we moved about a mile up the road and as the workshop was now a garage I decided to sell the Chinese milling machine and let the buyer strip it down and take it away, while the Tom Senior milling machine would be wheeled straight into the garage and be ready for immediate use.

I was planning to buy the same Chinese milling machine as I had before but was offered the Tom Senior for about the same money and I jumped at it. The Tom Senior is a more substantial milling machine, and I have not regretted the purchase.

Alterations

My Tom Senior milling machine was three-phase so I had to fit an inverter, but this was an easy job and only took a couple of hours including the forward/reverse switch and speed controller.

I also fitted a digital readout to the X and Y axes at the same time. I consider a digital

Fig. 2.5. The Tom Senior base cabinet with the central column fitted. The central column is very heavy even with everything removed. It will take three very strong men to lift it.

readout to be essential on a milling machine X and Y axes, but I can live without a readout on the Z axis although this could be useful at times.

The next thing I did was to fit a longitudinal X axis table feed. This is a desirable addition as you get a much finer finish than winding the table along by hand. Make sure you fit power-off switches at each end of the table; you do not want to burn out the power feed motor or ruin the leadscrew nut or even the leadscrew itself. *Model Engineers' Workshop* magazine has run several different articles on installing a power feed to a milling machine.

My Tom Senior has a couple of small cuts taken from the table by cutters that have pulled out of the collet (not done by me). This is not a major problem, but be sure that any cut-outs on the milling machine you buy

Fig. 2.6. This Tom Senior table has a couple of machining marks where the cutter has pulled down. Slight marks like this are acceptable on an otherwise good-condition milling machine, but if there are a lot of cutter or hammer marks on the table it may mean the machine has not been carefully looked after.

will not affect the accuracy of work or vices clamped to the milling machine's table.

Which collet system?

An important decision needs to be taken about the collet system used in the spindle. The smaller mills will have a 2 Morse taper or possible a 3 Morse taper spindle. A large turret milling machine would be much more likely to have an R8 taper. It seems to be easier to get second-hand 2 Morse taper collet holders than their 3 Morse taper equivalent. Better still would be to get the

Fig. 2.7. This boring head has a No.2 Morse taper shank. The top of the taper is threaded for a drawbar to stop it falling out of the spindle.

Fig. 2.8. The R8 taper is more likely to be found on the larger milling machines such as a VMC or Bridgeport turret milling machine. It is seen here on a flycutter.

R8 version if you can. R8 collets are much more substantial than 2 or 3 Morse taper shank cutters and, as they are often used in industry, more variations of R8 holders are available. You can even get a quick change R8 holder which is ideal for drilling holes as you can centre drill with one tool holder, drill with another and ream with a third. The holders are very quick to change; the holders only take a few seconds to swap over.

Whichever collet system you get, you must use a drawbar through the spindle. The collet must be held firmly and the cutter should be tight. Plain split collets are available in both 2 and 3 Morse tapers, but a better bet is to get an ER collet holder and a set of collets. These collets will close down in size, many by 1mm and the rest in ½mm steps. These will hold cutters very firmly and are the way to go. You do not need the entire set of collets at the start, just the ones that hold the standard shank cutters such as 6, 8, 10 and 12mm or their imperial equivalents.

There are various ranges available in the sizes we are likely to use: the ER32 range which takes up to a 20mm collet, the ER 25 range that takes up to 16mm collets, the ER 20 range which takes up to 13mm collets and the ER 16 range which takes up to 10 mm collets. I suggest you get the ER 20 range or the ER 16 range with a matching holder suitable for your lathe. The matching holder can be clamped down to the milling machine's table or to an angle plate to hold work to the milling machine. You can also add a 16mm side-lock holder to take the larger 16mm shank cutters and a 12mm side-lock holder if you need it. The side-lock holders will be more rigid than an ER setup on the smaller milling machine.

TESTING A SECOND-HAND MILLING MACHINE

The first thing I would do (after making sure it was safe to do so) would be to switch the milling machine on.

Earth leakage test

Before doing any further tests, to test a single-phase milling machine I would use an earth leakage trip. This is the type of earth leakage trip/tester sold for use with lawn mowers to stop you electrocuting yourself when you cut the lawnmower lead in half. This will not guarantee electrical safety but it will help to eliminate an unsafe milling machine at the outset. I would also use an electrical test meter to ensure the motor frame and the milling machine metalwork has a continuous earth.

DANGER LURKS

I recently heard about a lathe that had been supplied new with the electrical earth connection cut off. The owner had been using the lathe without an earth connection for seven years.

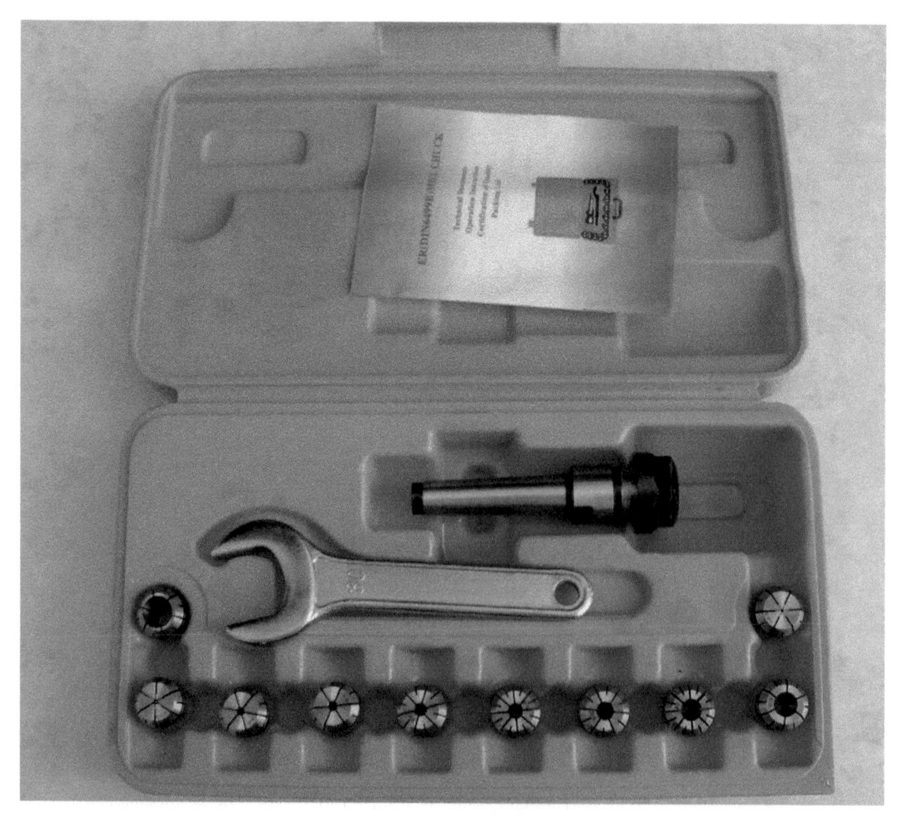

Fig. 2.9. This ER collet holder has a No.2 Morse taper shank. There are a wide range of collet sizes available.

Fig. 2.10. An electrical earth leakage trip can be used to check for electrical problems. Do not rely on this as a safety check; just use it as an initial check to eliminate some electrical problems.

Noise check

Next you need to run the milling machine in all its speeds from its slowest speed to its fastest speed and in back gear if it has any. You are listening for noise which indicates wear in the motor, belts, pulleys, spindle bearings and gears. You may be able to live with a noisy machine but can your neighbours?

Axis wear

Next, turn the machine off and wind each axis to the ends of their travel. Is there any wear? Do the slides feel loose in the middle and tight at the ends indicating the table has worn unevenly? Some unscrupulous people have been known to slacken the slides right off hoping that you will not notice the dovetail slide is worn in the middle.

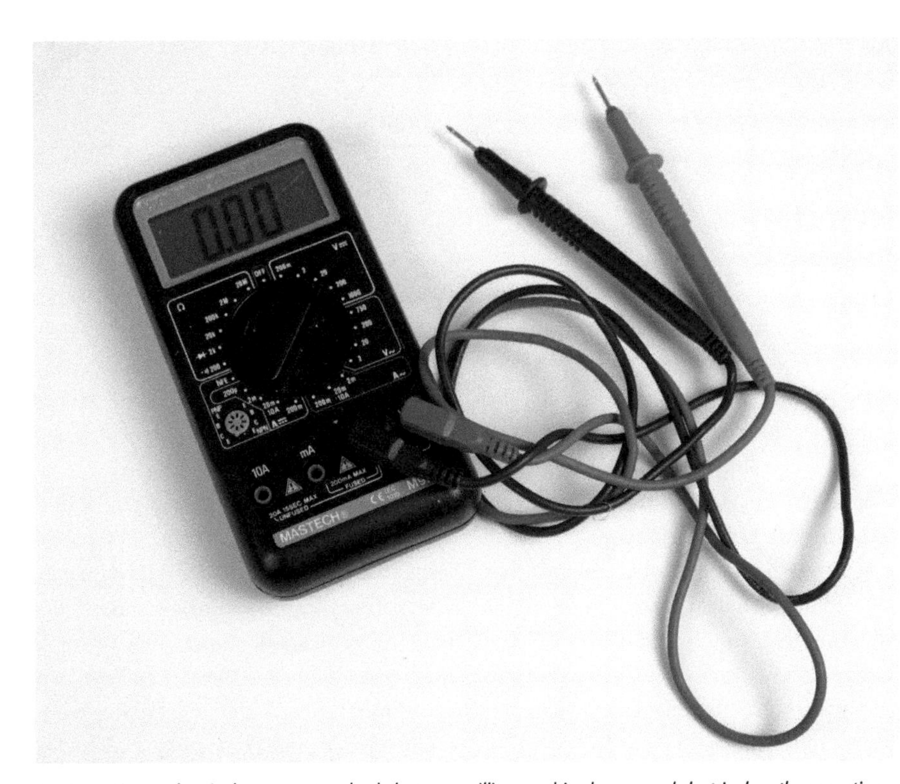

Fig. 2.11. Use an electrical test meter to check that your milling machine has a good electrical earth connection from the earth pin of the plug through to any metal components of the machine. (Note: this type of test meter is usually powered by a 9 volt battery, so it is not suitable for electrical resistance checks as the insulation may not break down until the test voltage is several hundred volts.)

Table parallel

The next test is most important. Put a dial test indicator into the milling machine's spindle. You may have to mount the indicator on a bar to extend it out from the milling machine spindle's nose. Wind the Y axis as close to the column as possible. Bring the indicator down on top of the milling machine's table as close to the front of the table as you can. Wind the Y axis towards you and watch the indicator. It should remain parallel within a thou if possible. Any more than this and I would not buy the machine, as the table is not square to the spindle and it would be a major job to correct this. Note: This is nothing to do with the squareness of the column; it is purely a check whether the table is parallel to the machine slides below.

TRANSPORTING THE MACHINE HOME

Compared with a lathe, a milling machine is often larger and heavier, and more often than not, it will be very top heavy. A small turret milling machine such as a Warco VMC or the Tom Senior Light Vertical milling machine can be stripped down for transportation, but the main body casting will be very heavy and will need two very strong men (preferably three or four) to lift it. However, if you have hard standing and good access, you can move the milling machine with an engine hoist.

If you do have to strip the milling machine down to move it, again an engine hoist will come in handy when you have to lift the main column back onto the base cabinet. If you do not have an engine hoist, there are two other methods that I have used to lift a very heavy milling machine main column onto its base. (Do not attempt either method on your own. Ideally two assistants should be available.)

(1) Stand the column vertically on the workshop floor. Then lift and slide it onto the top of the cabinet and then raise it upright; or

Fig. 2.12. An engine hoist will be found useful for lifting heavy parts to strip down and/or reassemble a milling machine.

(2) Stand the column in front of the cabinet. Then lean the column to one side and put a concrete breeze block under one side of the column; and then tilt the column the other way so it is raised by the height of the breeze block and put another breeze block under the column so the column is now raised by one level of breeze block. Repeat until the column base is level with the top of the machine base. Then you can carefully slide the column onto the base cabinet.

INSTALLING THE MILLING MACHINE

Bench or stand mounting

Hopefully, the milling machine you buy will come with its own maker's stand or even an integral base. A milling machine is a heavy item, so please ensure that any stand you buy or make for it is up to the task. Also, if putting the milling machine into a wooden shed, do make sure the floor is strong enough to take the weight. If in doubt, a thick sheet of MDF can be laid on the floor where the milling machine is going to stand. It is a good idea to cover the entire floor of the workshop with MDF anyway. It is easy to keep clean, and it will help distribute the weight and keep out the cold in winter.

If you are mounting the milling machine on a workbench, pay particular attention to the distribution of weight because a lot of workbenches have L-shaped legs made from thin sheet metal or angle iron and are liable to damage the floor. If in doubt about the weight loading onto the workbench legs, put some pieces of steel plate under the legs before mounting the milling machine onto the bench.

How to set up the milling machine and check for accuracy

The milling machine is fairly easy to set up accurately as it normally has a self-contained base that everything bolts onto. The milling machine's base will usually be bolted down to the milling machine's cabinet stand or bolted directly onto the workbench.

The normal hobby milling machine will have a round or dovetailed column that bolts onto the milling machine's base. It is important that this column is truly vertical. To check this verticality we will have to completely assemble the milling machine and check it for accuracy.

If you have purchased a new machine, you need to put a dial test indicator into the machine's spindle and check the table is flat as described previously for checking a second-hand machine. If this check finds that the table is not parallel, there will be no alternative but to get it collected by the manufacturer and a replacement milling machine delivered.

While it is important that the milling machine's spindle is square to the table, this can be sorted out on assembly, by scraping either the column mounting face or the base of the column mounting face. An alternative is to use some thin shim, paper often being enough to do the correction.

If the milling machine has a round column, a piece of round bar can be machined in the lathe and the column mounting casting machined true by clamping it on this diameter. Do support this mandrel with a running centre as the casting may be stuck well out. You may still have to scrape or shim the casting to the base, but you are starting from the correct position with the column square to the bottom of its mounting bracket.

MAINTAINING THE MILLING MACHINE

Oil and grease

Milling machines, like any other machine, require periodic lubrication. Try to use the correct lubricant as specified by the machine tool maker, or if the original is not available use a recommended substitute. The internet is useful for finding out the modern day equivalent of a discontinued lubricant or you could ask on the Model Engineer forum (http://www.model-engineer.co.uk).

There will probably be oilers dotted around the milling machine's sliding surfaces and you should top these up regularly. Check if the motor and any countershafts require lubrication. Most new milling machines will have sealed bearings, but some motors on older milling machines might have greasing points.

To lubricate the cutting tools, I suggest you use cutting oil rather than water-based soluble oil. The water-based oil will not protect the machine from rust or corrosion

Fig. 2.13. You should try to protect exposed slideways with a rubber cover to prevent swarf getting in.

over the long term and may even cause rusting. Oil and water should be kept well away from digital readout slides if fitted.

Adjusting the milling machine

All sliding surfaces should be adjusted to give minimum play while still allowing free movement. Time spent adjusting the milling machine will pay for itself with ease of use and minimum chatter.

Protecting the milling machine slideways

The slideways should be protected wherever possible. This can take the form of a piece of sheet rubber over the mill's sliding surfaces to stop swarf from falling onto them. It is easy to add the rubber; cut it to fit and sandwich the ends between a couple of strips of metal that can be screwed or clamped to the machine. You do not have to drill the machine; one end could be attached to the dovetail or round column with clamps, and the other end can be fixed to a round bar that just rests in the groove found around the machine's table.

FITTING A DIGITAL READOUT

A digital readout on any milling machine is essential. It will make life so much easier and will help to eliminate machining errors as well as saving on that most useful workshop resource – time. It is quite easy to fit a digital readout to a milling machine; you just need to be accurate and logical in your approach.

I fitted the X axis scale to the outside of the table on the Tom Senior Light vertical milling machine used to illustrate this book. If I had put it on the inside edge of the table, I would have reduced the Y axis travel by about 30mm, clearly not a good idea. (Modern magnetic digital scales are thin enough so that you can mount them on the back of the milling machine's table with a minimum loss of table travel.)

X axis scale

The first job was to fit the scale to the front of the table, close to the top edge but not above. I clamped the scale's back plate to the front of the table with a 1/8in packing piece between the back plate and the table. You may need to use a different thickness of packing depending on your milling machine. The back plate was set parallel to the table using a height gauge and an old slip gauge, although a parallel piece of material and a dial test indicator on a scribing block would do just as well.

Fig. 2.14. Fitting the readout scale's back plate to the front of the milling machine's table. The backplate must be true to the top of the machine table. I used clamps to ensure the back plate was held on securely.

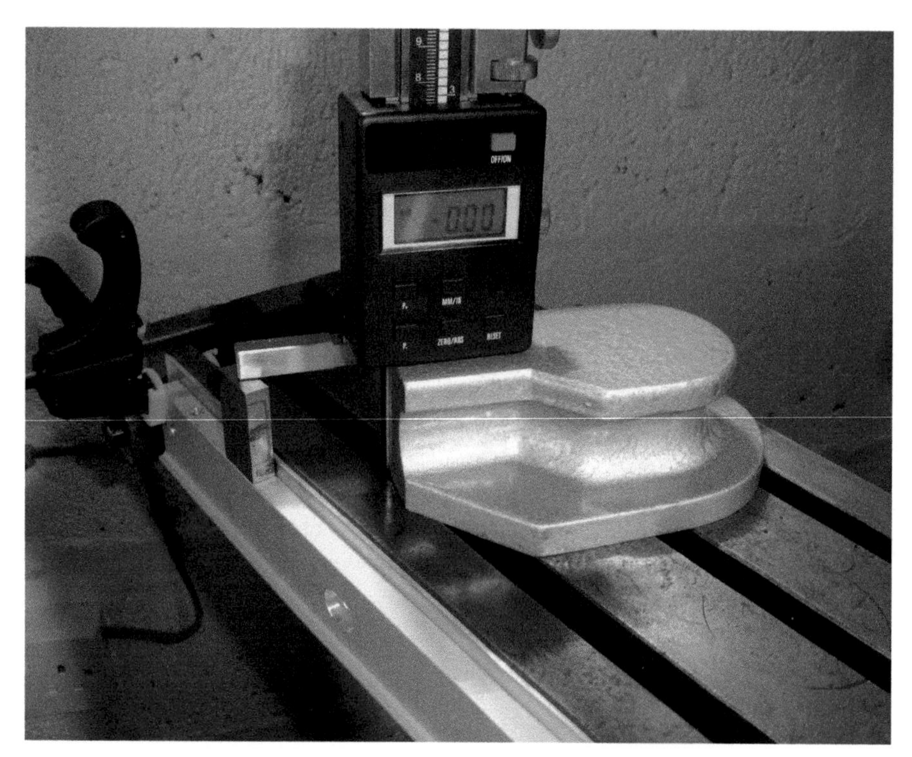

Fig. 2.15. *A close-up of the backplate. Check for zero at each end of the bar.*

Fig. 2.16. *A typical transfer punch. This one is spring loaded. In use the cone is pushed into the hole, and then the end of the punch is given a light tap with a hammer to make a centre punch mark.*

I used a transfer punch to mark the machine table through the back plate. Then I drilled and tapped the milling machine's table to fix the back plate on. I checked that the back plate was parallel to the top of the table by using a height gauge and a parallel to ensure all was correct.

I held the scale onto the back plate and moved the reading head to a suitable position in the middle of the cross slide casting. Do not put it at the end of the casting as we do not want the reading head to foul the ends of the readout scale. I again used the transfer punch to mark the mounting holes for the scale and the reading head.

The reading head should remain in alignment in all directions – horizontally and parallel to the table – and not interfere with the table's oiler or travel lock. The reader will probably be fitted with a couple of plastic alignment clips to hold the reading head in the correct position. Do not remove these clips until the scale and reader head are all screwed up tight. (The X traverse locking screw may need replacing with a longer screw to operate.) Finally, fit the cover and the X axis scale is finished.

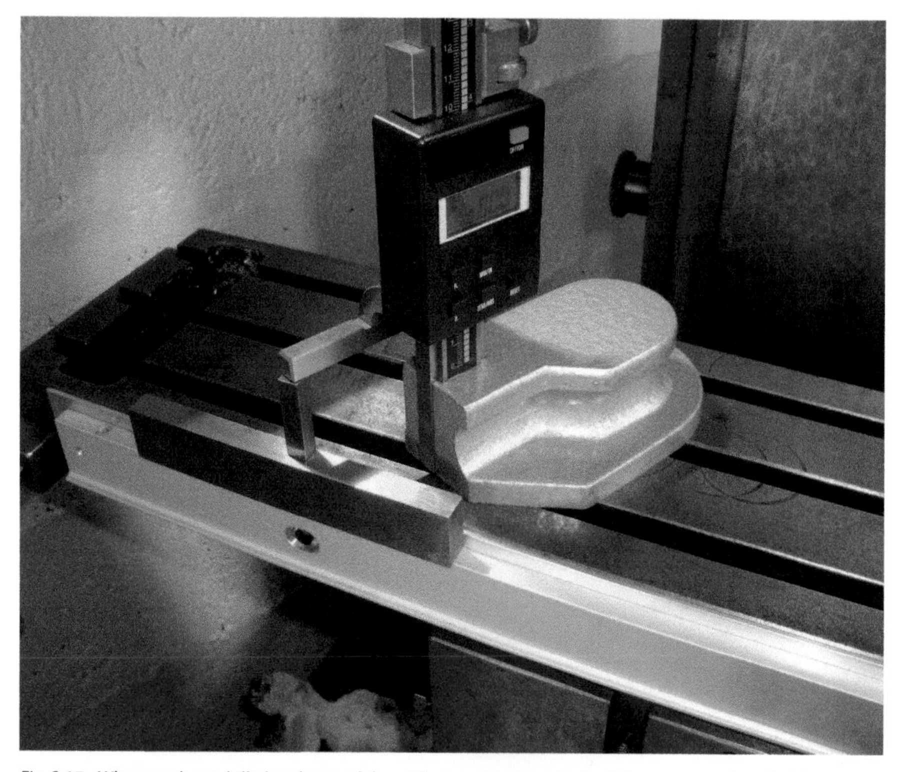

Fig. 2.17. *When you have drilled and tapped the milling machine's table, check for zero at each end of the bar. Then tighten the screws right up as tight as you can and make sure the back plate has not moved while tightening up.*

Fig. 2.18. Positioning the readout head to the front of the milling machine's knee casting. You may need a spacer or two behind the readout head.

Fig. 2.19. The readout head and the scale have been fitted.

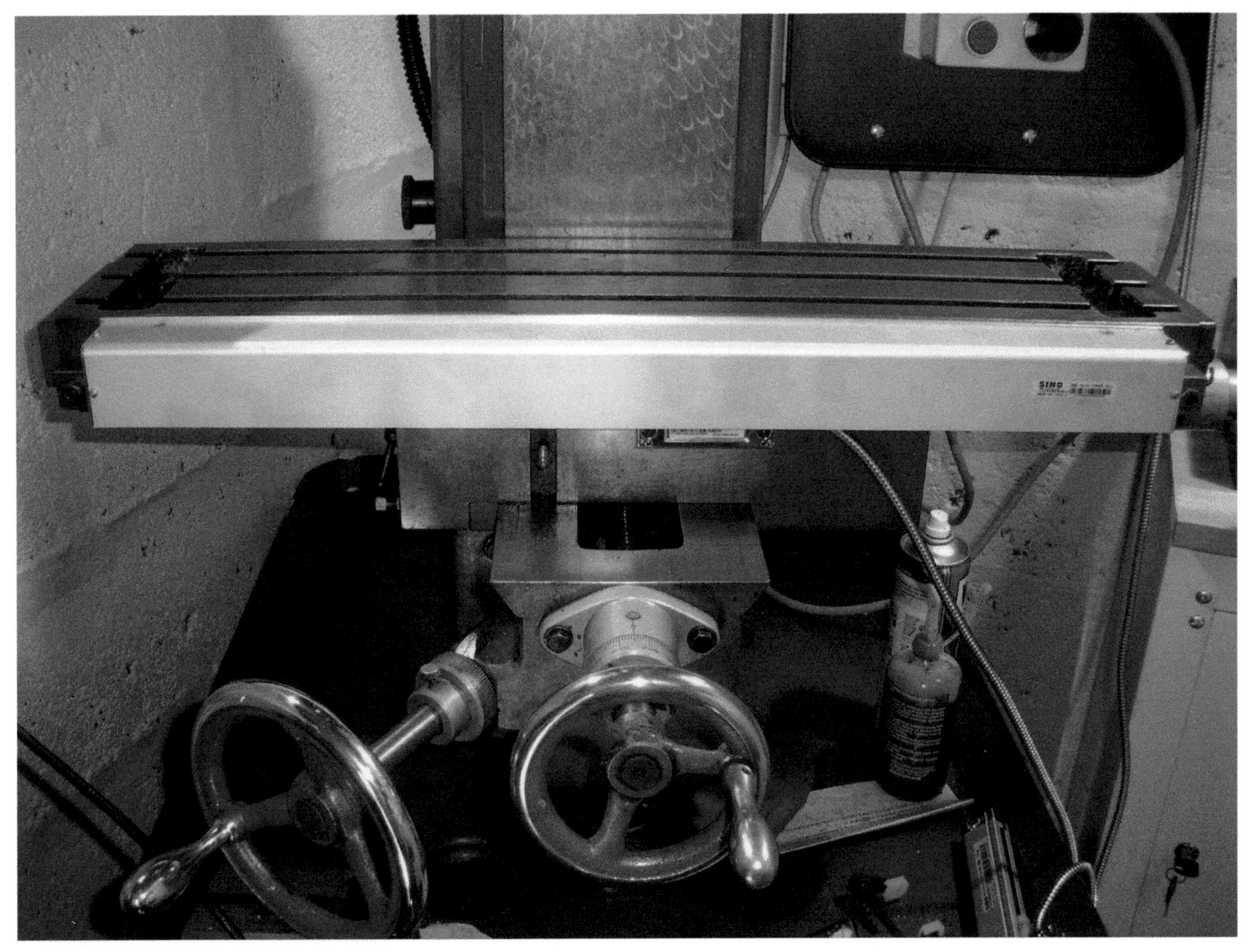

Fig. 2.20. The finished X axis scale with the cover fitted.

Fig. 2.21. The Y axis backplate being fixed to a piece of 2×2in aluminium angle.

Y axis scale

Now we move onto the Y axis. I mounted a piece of aluminium angle to the Y axis back plate of the readout slide ensuring it was parallel to the edge using a surface plate (use the milling machine table if you do not have a surface plate).

I fixed the angle, together with the back plate, to the underside of the cross slide and clocked it true. It is very important that the angle is parallel to the machine's Y axis travel in both directions.

Fig. 2.22. The Y axis backplate and mounting angle being clocked true.

Fig. 2.23. The Y axis scale has been fitted and the readout head has been screwed to a bit of cut-down 2×2in angle. One of the brackets supplied with the readout has also been bolted to the knee casting.

Fig. 2.24. The readout head bracket and the knee bracket have been clamped together.

Fig. 2.25. The two brackets are finally bolted together and both scales are now fitted.

The scale was fitted next and the reading head was screwed to another bit of aluminium angle. A large aluminium bracket was fixed to the knee casting and was also fixed to the bottom of the angle that the reading head was mounted on. This bracket was free to move up and down to align it with the reading head bracket. All was then tightened up and both scales are now fitted.

Reading head clips

Remove the red reading head clips to free up the reading heads. These are the red clips held in by a screw at each side of the reading head. Put these clips and screws away safely in case you ever need to remove the scales, perhaps when moving house. You will need the clips to refit the scales. One of the brackets I used was supplied in the readout's kit of parts and some I made from $2 \times 2 \times \frac{1}{4}$in aluminium angle.

Finally, the digital display is mounted onto the side of the machine. You could mount this on the wall near the machine if the machine is not big enough. The readout should be fitted on a swivel so you can move it towards you when it is needed and out of the way when not required. Once you have used a digital readout, you will never want to go back to using the handwheel graduations.

Fig. 2.26. The digital readout mounted on the side of the mill.

3 Using the Mill Safely

Most of you reading this book will probably be working alone in a small workshop, your garage or a garden shed. If you get hurt, you are possibly on your own. As a priority, I suggest you take a walkabout or mobile phone into the workshop with you and keep it in your pocket. The ability to quickly contact your family, a friend or the emergency services could save your life. Do not attempt to work on a milling machine and talk to someone at the same time. If you have to talk to someone or answer the telephone, switch the milling machine off before doing so.

The area around a milling machine should be kept free of obstructions and the floor should be clean and dry. Any oil or other spills should be cleaned up immediately. The milling machine should be kept clean and tidy.

Workbenches placed near to a milling

Fig. 3.1. A wooden shed can be used as a home workshop.

Fig. 3.2. If you have one, a brick-built shed will make a good workshop.

Fig. 3.3. The inside of the brick-built shed. Notice the painted floor and the rubber mat. You can buy proper floor paint at your local DIY superstore. Although this shed is quite small, 10×5ft internal dimensions, I have fitted in a large mill, a Myford ML7R lathe, a drilling machine and a cabinet to store milling accessories. Do try to keep the workshop clean and tidy, as it helps ensure safety.

machine should be strong and equipped with a non-slip safety surface to stop items such as tools from rolling onto the floor – a piece of rubber car mat will suffice. Measuring equipment, tooling and workpieces should not be left on the milling machine table as they can vibrate off onto the floor.

To save you getting swarf embedded into the bottom of your shoes a rubber door mat on the floor will allow swarf to drop through.

REMEMBER, YOU ARE RESPONSIBLE FOR YOUR OWN SAFETY

Be mentally alert while using the milling machine. Avoid operations that would put you at risk of your hands getting caught in the cutter. Do not operate any milling machine when tired or while under the influence of drugs, alcohol or prescription medication that may cause drowsiness or otherwise affect your ability to control the milling machine safely.

USING THE MACHINE

Starting up

Always double check to make sure your work is securely clamped in the vice before starting the mill. Rotate large cutters such as face mills or flycutters by hand to make sure they clear the work. Be certain that the cutter turns freely and is firmly mounted. *The speed of the milling machine must be checked before turning it on.* If necessary, start the milling machine at a low speed and increase the speed gradually.

Stopping the machine

Do not use any milling machine before you are familiar with the controls and are aware of the ways that you have of stopping the machine. This may be a clutch, an on/off switch, a No Volts release switch or an emergency stop button. If your milling machine does not have a No Volts release switch, I strongly advise you to fit one. As well as being useful to stop the milling machine in an emergency, it will also prevent the milling machine from starting up by itself should a power cut have occurred. If you are not comfortable working with electrical equipment, get a qualified electrician to fit the switch for you.

Shut off the power supply to the motor before doing any operations that require access to the mill, such as changing tools, measuring work, mounting or removing accessories and so on. The milling machine must be safely stopped. Just zeroing a

Fig. 3.4. You should fit and use a No Volts release switch on any machine in the home workshop.

variable-speed drive is not safely stopping it. Unplug the machine, switch it off and use the No Volts switch as well.

Leaving the mill

Never ever walk away, even for a moment, while the milling machine you are using is switched on. Switch it off if you are leaving it unattended, and only leave the milling machine when it has come to a complete stop.

Never leave a milling machine in an unsafe condition. The chances are that if you are halfway through setting up the milling machine, when you come back you will forget what you have or have not done. If in doubt, leave a big notice for yourself telling you to check that the milling machine is safe before continuing. A whiteboard with suitable felt-tip markers is useful for this.

GUARDS

New milling machines will be fitted with guards over the motor, the drive belt and the spindle pulleys. These guards should always be in place before using the mill. Any other moving parts should be guarded for safety. Guards should be made from suitable material and not have sharp edges.

Guards are often removed by machinists, but if the guards are constructed from clear materials and open and close easily, they add safety but do not interfere with the work being done. The guard will, at the same time, also deflect swarf and any coolant that is being used.

Guards fitted to the milling machine should not be removed while it is in use. Unless unavoidable they should not be removed for adjustment and lubrication of the milling machine. (Please note that guards are not including in the photographs in this book to make for ease of photography.)

Chip guards

Milling machine manufacturers do not usually supply chip guards, so adding one should be considered. Chip guards with magnetic bases are available, although they can attract steel swarf to the magnet so put the magnetic base into a plastic bag to aid swarf removal.

A simple home-made guard for the milling machine can be made from a piece of plywood or similar. It just needs to be a rectangular shape and simply rests in the front T-slot. Several home-made guards of different heights to suit the job should be available. Add a cut-out in the middle of the guard to fit over the body of the machine vice as necessary. It will protect you from flying swarf, and from swarf spinning around when you are drilling holes. It takes seconds to fit and seconds to remove. Please do make some and use them.

CLOTHING AND EQUIPMENT

Protective gear

Milling machines can throw sharp metal chips considerable distances, especially at high cutting speeds. Apart from deep cuts, the chips may be red hot when they leave the tool and can cause skin burns and even burn holes in clothing.

Always use eye protection; wear industrial quality safety glasses fitted with side-shields. Glasses without safety lenses are not adequate eye protection. When the safety glasses get dirty, do not wipe them off as they will become scratched – wash them off under the tap instead.

Do not attempt to remove chips or metal swarf with your hands; always switch the milling machine off first and either pull the swarf away with a large metal hook or lift it off using a pair of gloves. Hand gloves should never be worn while operating a mill,

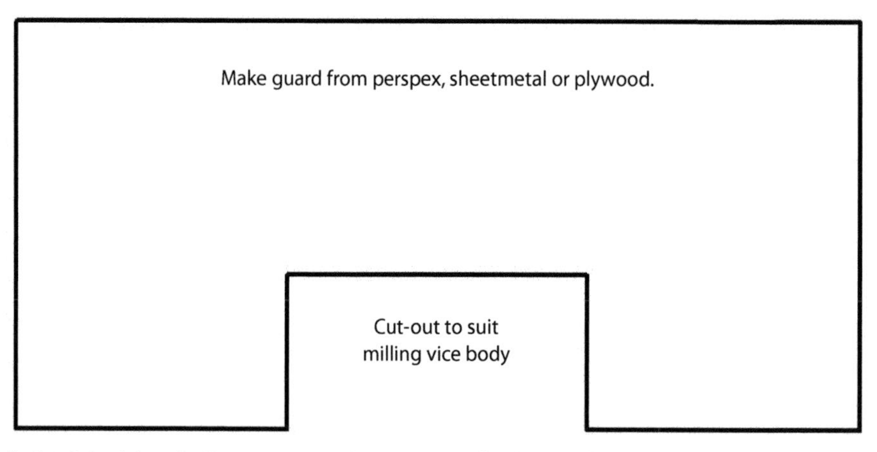

Fig. 3.5 A simple board with a cut-out in it will do to stop swarf and coolant flying around. You can make it quite easily from a bit of sheet metal or Perspex.

and while protective gloves may be worn when swarf is being collected, remove them before switching the milling machine back on. If the swarf is tangled around the spindle or work, untangle it with a pair of pliers, not a gloved hand; it will cut straight through the gloves and into your hand. Use a dustpan and brush to remove small chips and swarf from the swarf tray – never use your hands.

Swarf can get inside protective clothing, especially collars, pockets and shoes (use a clothes peg to keep your collar tight).

Wear steel toe cap safety boots; they are not that expensive and they will help to protect your feet from being injured if you drop a heavy object on them. The 'trainer' version is particularly comfortable.

Other clothing

If overalls are not worn, long shirt sleeves should be rolled up above the elbows or better still, wear short sleeve shirts. Loose sleeves can catch on rotating work and quickly pull your hand or arm into the cutter. Obviously, if hot swarf is flying around, keep your arms protected.

Under no circumstances wear a necktie when operating a mill.

Long hair should be tied back; if it gets caught in the rotating spindle, you could be pulled into the mill.

Do not wear wristwatches, rings or jewellery when in the workshop as they could catch on a cutter. The same goes if you are moving a milling machine into or out of the workshop; they could catch on a falling mill.

SAFE PRACTICE

Lifting

Back injuries can be caused by incorrect lifting practices. The important thing to do is to bend your legs to lift rather than bending your back. A hoist or engine lift should be used to lift heavy items. A stool in the workshop will take pressure off your legs and will make a welcome break after you have stood at the milling machine for a while.

In case of fire

If you are machining titanium or magnesium both of them can ignite and burn vigorously so keep the accumulation of swarf to a minimum. If the material does catch fire, don't use water or water-based coolant as an extinguisher, it will make matters much worse. Before machining, get a large bucket and fill it with dry sand, cover it and keep it by the mill. The sand will smother the flames very effectively. You could also buy one of the special fire retardants used in industry.

Lighting

The workplace should be provided with uniform lighting that gives an adequate level of illumination. Good lighting will help you to produce good work. I have read elsewhere that fluorescent lights can make the machine appear static due to the stroboscopic effect of the light on the machine's rotation. I have never had this happen to me but unless you are deaf, you will be able to hear if the machine is running as well as see it running so I would not worry about this problem occurring.

MIND YOUR SPEED

The milling machine must be set to operate at the speeds and feeds recommended for the specific metal being milled. If the cutter rotates too slowly/fast or the feed is too slow/fast, broken cutters and/or accidents are likely to occur.

Flycutters

Flycutters can be particularly dangerous. A flycutter is usually quite a large diameter and has a single cutting tooth, often extending

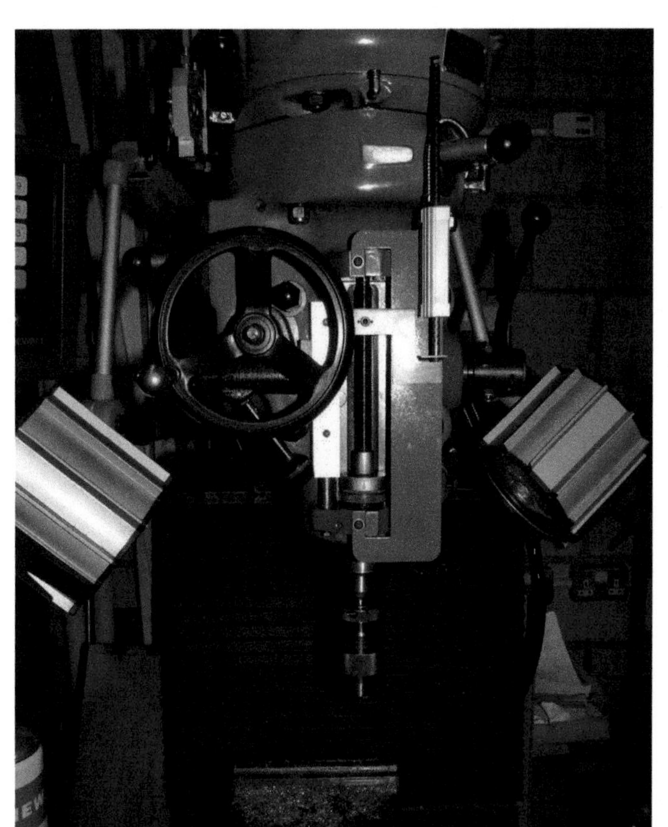

Fig. 3.6. Low-voltage machine lights are easily fitted and make it easier to see the work. Caution: lights do get hot in use; try not to touch them when switched on and do not let soluble materials come in contact with hot surfaces, especially the glass bulb protectors.

Fig. 3.7. A single-tooth flycutter will sweep quite a large surface with little loading on the mill's motor. Keep your hands well away from the sharp cutting edge.

outside the main body of the flycutter. As flycutters are usually run at a fairly high speed, this cutting tooth is often running so fast that it is virtually invisible. Also, the swarf will fly considerable distances. So, when running a flycutter keep your hands well away from the cutting tooth, always use a guard, and always make sure the swarf is flying away from you, not towards you. I have never had a serious accident in the workshop, but the closest I came to it was with a flycutter, it just nicked a slice of skin off a finger because I put my finger too close after switching the machine off. The cutter had not quite stopped, it could have been much worse. You have been warned.

UP-CUT AND DOWN-CUT MILLING – A WARNING!

There are two types of milling cut, up-cut and down-cut. Up-cut milling pushes the material and the job away from the cutter. Down-cut milling pulls the work into the cutter. On simple manual milling machines like we are looking at here, up-cut milling is the only method we shall be using.

Down-cut milling could cause a nasty accident by pulling the work into the machine, possibly shattering the cutter and damaging the milling machine and even damaging you. CNC mills and machining centres have special leadscrews suitable for down-cut milling.

Just because you see down-cut milling elsewhere does not mean you should use it.

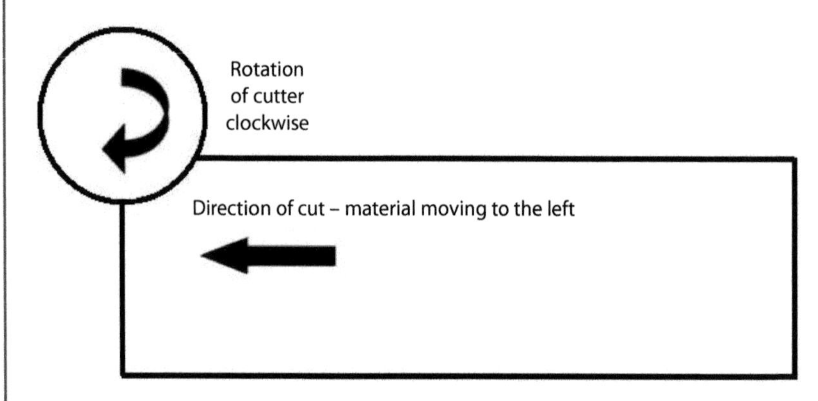

Fig. 3.8 Down-cut milling will tend to pull the cutter into the work with dire consequences. Down-cut milling is normally OK on a CNC machine that has ball screws on the lead screws.

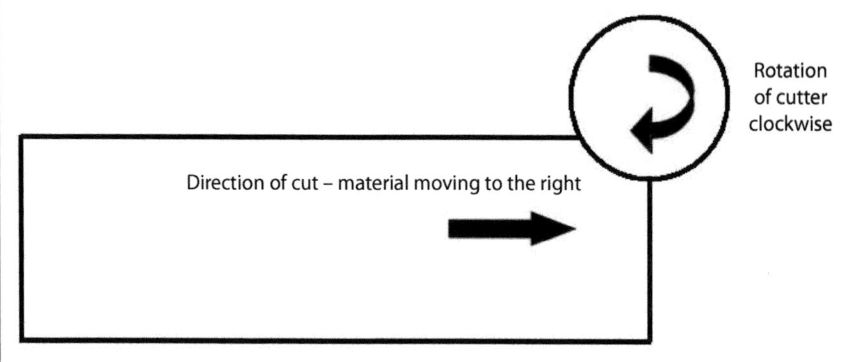

Fig. 3.9 Up-cut milling tends to push the work away from the cutter. This is the normal method you should use on an ordinary home workshop milling machine.

4 Cutting Tool Principles

Cutting tools such as slot drills and end mills are available in high speed steel (HSS) and tungsten carbide. The tungsten carbide milling cutters can be either ground from solid tungsten carbide or made with a toughened steel body that takes inserted tungsten carbide inserts. Steel shank cutters with brazed-on carbide inserts are also available.

CUTTING TOOLS

Slot drills

As their name suggests, slot drills are mainly used for cutting slots. This can be as simple as a keyway or it can be a complicated shape which bears no resemblance to a parallel slot. They can also be used for machining a pre-drilled hole to size. Most slot drills can be used to plunge drill straight into a solid chunk of metal as the two flutes are usually ground off-centre. This type of slot drill is usually known as a centre cutting slot drill.

Fig. 4.1. A standard slot drill normally has two flutes. Three-flute slot drills, normally called supermills, are available and are ideal for the fast removal of aluminium. This is a standard slot drill cutting a small chamfer on a component.

Slot drills are also available with three flutes, and these again are usually centre cutting. The three-flute slot drills are used in a similar way to two-flute slot drills and are often designed to remove soft metals like aluminium at a very fast rate.

End mills

End mills are similar to slot drills but usually have four flutes, although six, eight and even more flutes are available on some larger cutters. The end mill normally has a centre in each end which is used to manufacture the cutter and to resharpen it. This means the end of the cutter is hollow and so it cannot be used as a centre cutting cutter. However, centre cutting end mills are now available and can be used like a slot drill.

However, you can mill on both the sides and the end of an end mill even though you may not be able plunge a hole with it. End mills are available in various helix angles from straight (for brass) to a very high helix for removing harder materials while still leaving a good finish.

You should not attempt to cut a slot with an end mill; the most material you should try to remove with an end mill is the diameter of the end mill divided by four, and the

Fig. 4.2. An end mill of the size used in the home workshop normally has four cutting teeth. Larger end mills can have six, eight or even more teeth. The cutter shown here is a standard series cutter. Long series cutters have a longer cutting edge than a standard end mill.

Fig. 4.3. This end mill is centre cutting so it can be used to plunge cut a hole if necessary.

Fig. 4.4. A diamond hone can be used to chamfer the corner on a cutter or to hone a small radius on the corner.

depth of cut should be equal to or less than the diameter of the cutter.

End mills are not designed for facing, but if you do have an old end mill in need of sharpening, you can stone (or diamond hone) 45-degree chamfers on the corners which will give you a better finish when facing.

FC3 cutters

FC3 cutters are commonly known as throwaway cutters as they are designed to be used and then thrown away when worn out. Their cost is so small that, at least in industry, it is not worth the expense of sharpening them. As their name suggests, they normally have three flutes. They are available in many sizes although their shank is normally ¼in (6mm) in diameter. I have seen FC3 cutters with larger shanks such as 8, 10 or 12mm.

FC3 cutters have a parallel shank, and normally they have a flat on the side of the shank for a grub screw. FC3 cutters are quite sharp, but I have found that many makes of FC3 are not quite sharp enough for use on plastics so I would stick to normal slot drills for plastics.

You can use FC3 cutters to cut narrow keyways in shafts to take a standard key. If used with care, they will cut a keyway which is about 1 thou narrower than the nominal key or cutter size. You can plunge at one end, wind the cutter along by the length of the keyway required and then raise the cutter from the slot. The key will probably need filing down very slightly to fit the keyway.

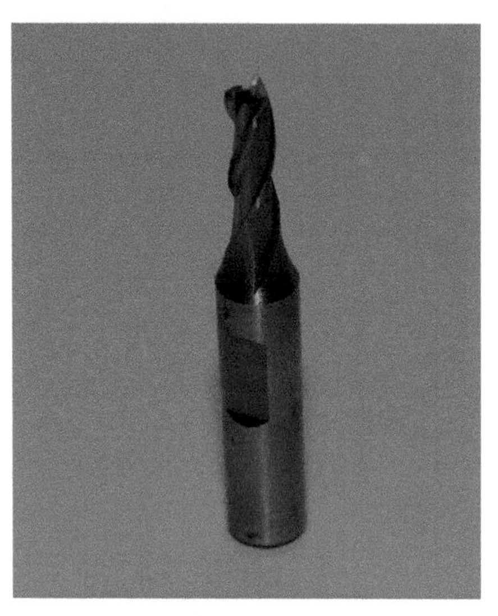

Fig. 4.5. Small three-flute cutters, called FC3s, are useful for cutting small slots and keyways. They usually have a 6mm or ¼in shank although 8mm, 10mm and even 12mm diameter shanks are available. The FC3 cutter is normally regarded as a throwaway cutter in industry, but it is possible to resharpen them in the home workshop. I find FC3 cutters are not normally sharp enough for use on profiling plastic components.

Roughing cutters

Roughing cutters are used to remove large amounts of material in the shortest possible time. The edge of a roughing cutter is most often of a sinusoidal shape although straight edge roughing cutters are available. These are similar to normal cutters but they have a series of staggered nicks all round to break the chips into smaller lengths.

I believe the original roughing cutter was the Rippa cutter made by Clarkson with a very coarse pitch. If you imagine a cutter with a coarse sinusoidal thread along it with flutes, you can see it will remove a lot of material in a short space of time. Over the years, the Rippa cutter has been refined by other makers, and various pitches of roughing cutters – fine, medium and coarse pitch – are now readily available.

The fine roughing cutters are very good on the tougher materials such as stainless steels and some of the more exotic alloys, while the medium roughing cutters are good for steel. The coarse ones are excellent at removing material from blocks of aluminium, but do use some cutting lubricant as the aluminium will start to stick to the teeth of the cutter.

Shell mills

Shell mills are not very common nowadays but you can pick them up second-hand. Shell mills are like very large multi-fluted end mills and are usually mounted on a separate shank. They are ideal for facing off large bars or castings.

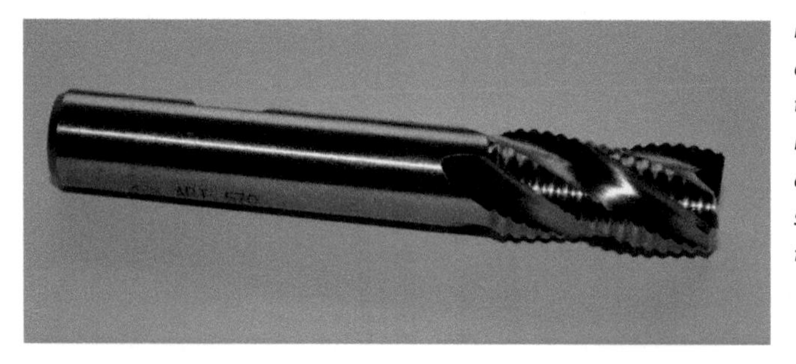

Fig. 4.6. Roughing cutters are designed to remove the maximum amount of materials in the minimum amount of time. The cutter shown here is a medium tooth cutter.

CORNER ROUNDING CUTTERS

Corner rounding cutters, as their name indicates, are used to round the corner of a block of metal or similar. They usually have four straight flutes with a ready-ground radius on them. They are easy to set: lower the bottom of the cutter to the top of the work, then raise the cutter by the radius you are milling. To use, wind the cutter diameter to the side of the work and move over by the radius; then when you run the cutter along the work, you will machine a radius all along the edge of the work.

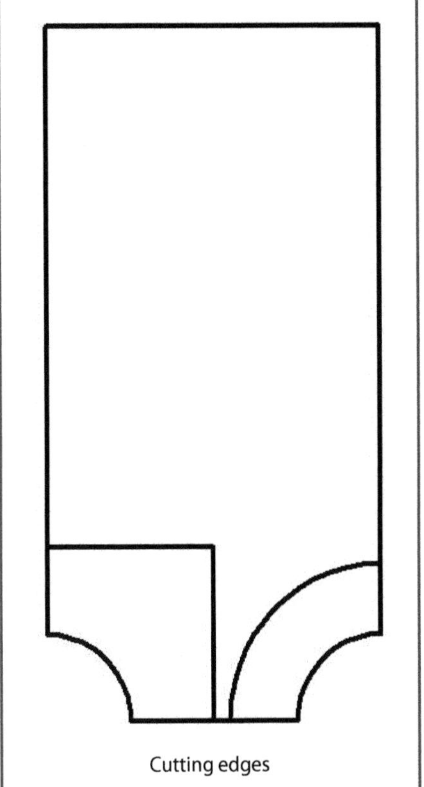

Fig. 4.7 A corner rounding cutter will put a radius on the edge of a component.

Fig. 4.8. Using a ball end cutter to cut a bearing slot in a casting.

Fig. 4.9. The bearing in the finished slot. A gap of 0.005 of an inch has been left all around to allow for the paint.

Ball nose cutters

The ball nose (or bull nose) cutter is ground to the shape of half a ball on the end. It can be used to machine grooves with a semi-circular bottom along a bar. It can also be used to profile work.

T-slot cutter

A T-slot cutter is designed to cut T-slots. It has staggered teeth and it will cut on the diameter, the top and the bottom. Before using a T-slot cutter, you should cut the basic slot to the required width and depth, then cut the T part of the slot with the cutter.

Woodruff cutter

A Woodruff cutter is used to cut a semi-circular slot in a shaft to take a semi-circular Woodruff key. The Woodruff cutter only cuts on its outside diameter.

Dovetail cutter

A dovetail cutter is used to cut dovetails, most often in cast iron for machine tool slides. You can get dovetail cutters in both 45-degree and 60-degree versions. You can also get reverse dovetail cutters that will cut an inverted dovetail.

Flycutters

Flycutters are useful for all sorts of tasks, although they are most often used to face large blocks of metal. For steel and most other metals, the flycutter should be sharpened with a small leading chamfer. However, for plastics, the flycutter should be sharpened with a leading shallow point. This will slice the top of the plastic and leave a polished finish rather than a torn finish. Flycutters can also be used with a shaped tool bit to machine various different shapes on the component being machined.

Fig. 4.10. Flycutters will clean up large faces in one sweep.

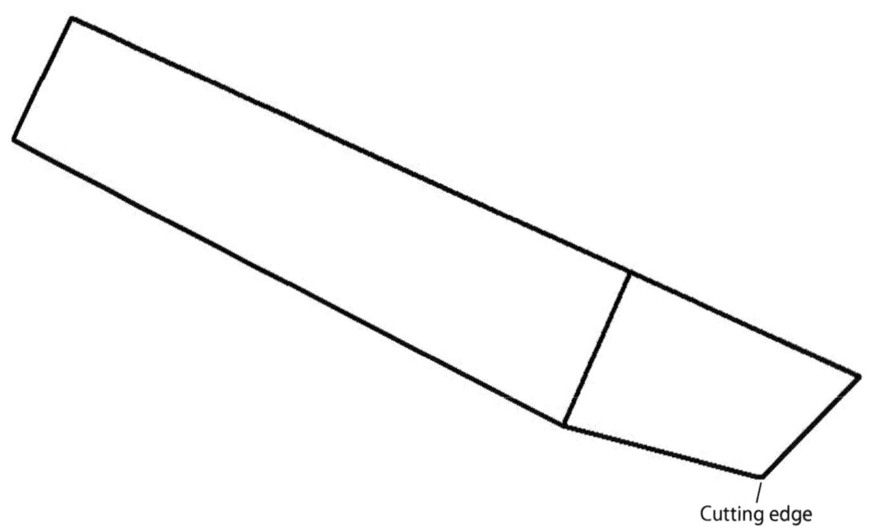

Fig. 4.11 This flycutter shape is ideal for metals such as steel, aluminium and brass.

Slitting saws

Slitting saws are for cutting slots or for cutting items into two. Slitting saws are available in various widths, diameters and tooth pitches. Be careful when using a slitting saw as the material being cut may close in on the saw and trap it as you take the cut deeper. You could always add a piece of round bar inside the hole being slit to stop the work closing up onto the bar. Do make sure you restrain the round bar in some way so that it cannot move while the slot is being cut.

There are small slitting saws available which are ideal for cutting the slots into screws for projects such as clock making. The normal slitting saw is hollow ground and has no set on the teeth so it can only cut on its diameter.

Fig. 4.12. A slitting saw is used to cut screwdriver slots.

Chipping saws

Chipping saws are like slitting saws but they have side cutting teeth, although they are still designed to cut on the diameter. While they should not be used to cut on their sides, the side teeth will help to remove the swarf so they are much less liable to clog up and seize.

Side and face milling cutters

These cutters are like chipping saws but are normally wider and stronger. They can cut on the sides as well as the diameter and

Fig. 4.13. A side and face cutter.

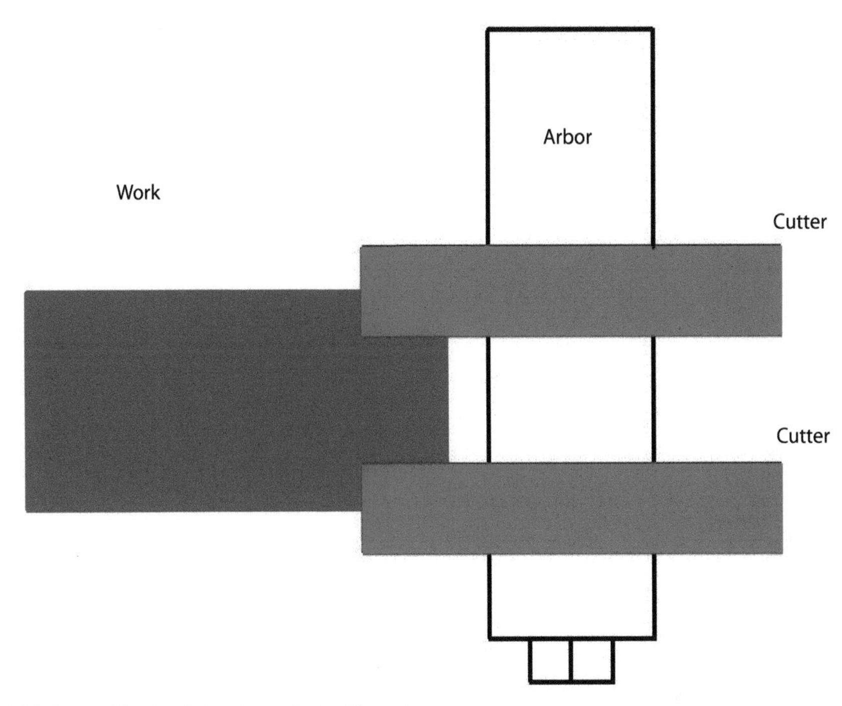

Fig. 4.14 Gang milling two flats with a pair of milling cutters.

would most often be used to cut a wide slot right through a piece of metal, this often being done in more than one pass to make a slot of a specific width. As they are quite thick, they are unlikely to deflect.

Side and face cutters are also available in what are termed staggered tooth cutters. Every other tooth is cut away on the sides so that the cutters can interlock with each other. This means that you can insert shims in between the cutters to space them out. If you insert the correct thickness of shim, the pair (known as a gang) of cutters can be made to cut a specific width of slot. This is often used in a production environment so the slot remains a constant width. You can also shim the cutters out so that they will mill a specific width such as a flat either side of a round component. Milling both sides at the same time saves a machining operation and cuts the machining and handling time in half.

Gear cutters

Gear cutters are for cutting the teeth on gear wheels. A normal set of gear cutters contains eight different profiles and are numbered 1 to 8.

Cutter No.1 cuts gears of 135 tooth or more, including racks.

Cutter No.2 cuts 55–134 teeth.

Cutter No.3 cuts 35–54 teeth.

Cutter No.4 cuts 26–34 teeth.

Cutter No.5 cuts 21–25 teeth.

Cutter No.6 cuts 17–20 teeth.

Cutter No.7 cuts 14–16 teeth.

Cutter No.8 cuts 12 and 13 teeth.

Gear cutters are normally of involute form, and Imperial cutters are designated by the diametral pitch (DP) or the metric equivalent which is the module.

Fig. 4.15. A set of eight gear cutters.

Hole saws

Hole saws are designed to cut holes in sheet metal. They are made in the form of a tube with teeth and fit on an arbor with a pilot drill. They are made in many different sizes.

Rota broach

A rota broach is similar to a hole saw but does not usually have a pilot drill. They will cut a circular hole and are often used on thicker material than a hole saw. They are also useful for cutting through tubes that need a scallop cut out of them before welding up.

INSERTED TOOTH CUTTERS

Small cutters and face mills

Inserted tooth cutters can be of quite small diameters with one or two inserted teeth, or several inches or centimetres in diameter taking many inserted teeth. The latter are known as face mills as their main use is milling the face of blocks of metal, either to clean up the face or to remove large quantities of metal quickly. The face mills are also useful for machining the face of castings to remove the hard skin and clean up the face.

The smaller-diameter inserted tooth cutters are more likely to have an integral shank, while much bigger face mills normally have a separate shank with the actual face mill bolted on. Milling cutters (and the separate shanks for face mills) are available

Fig. 4.17. This inserted tip has eight sides. They last a lot longer than the more usual three or four sided tips.

in a wide range of diameters and different lengths. The eight-edge octagonal shaped inserted tip puts a much lower load onto the milling machine than some other types of chip.

Coatings to reduce tool wear

Inserted milling tips are also made with different coatings that improve tip life. One of the most common coatings is TiN (titanium nitride), which is gold in colour. Some other coatings offer slightly better wear resistance; they are TiCN (titanium carbon nitride), which is a blue grey in colour and TiAlN (titanium aluminium nitride), which is a violet bronze.

Inserted tip tools need a greater depth of cut, a higher speed and a higher feed than HSS tools. Because of this, to take a heavy cut, the milling machine needs to be rigid and have ample power to drive the cutter; this rules out using tungsten carbide tools on the smaller mini-mills. These smaller machines just do not have enough speed or power, so most likely the tool will chip.

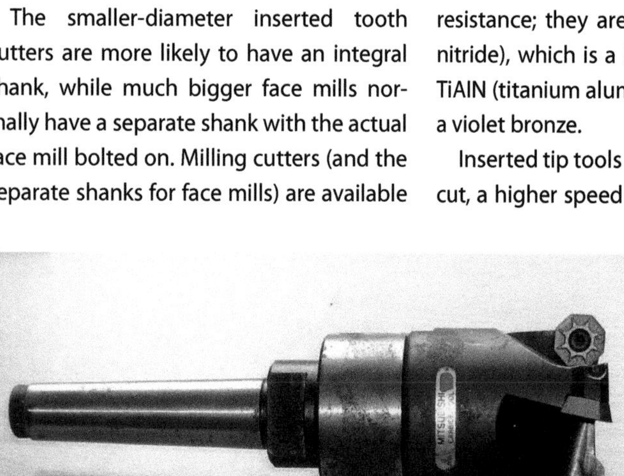

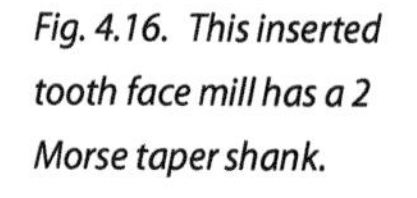

Fig. 4.16. This inserted tooth face mill has a 2 Morse taper shank.

Inserted tip types

There are many shapes and sizes of inserts. Some milling inserts are made to an international standard while others are made to the maker's own standard. Wherever possible, use ISO tips and cutters as they will be interchangeable with other manufacturers. The USA has its own system, the ANSI system, which is not compatible with the ISO system as used in the rest of the world.

CUTTER HOLDERS

There are a few systems of holding cutters, standard end mills and slot drills in the milling machine. You can get the inserted tip cutters with both plain parallel shanks and Morse taper shanks. Most end mills and slot drills come with parallel shanks, sometimes with a cut-out to take a grub screw in a side-lock holder.

Side-lock holders

A side-lock holder has an outside taper or shape (usually 2 or 3 Morse taper or R8) to fit the machine.

The holder will have a precision ground bore (such as ½in (12mm) that is a good fit on the cutter. The cutter will be tightened up by a grub screw which presses on a parallel flat on the side of the cutter shank. Ideally the holder will have an internal grub screw or stop so the cutter cannot push back in the chuck. There is another type of side-lock holder, where the cutter has a flat ground on a slight angle (usually a couple of degrees) and deeper at the lower end, so the cutter cannot pull out of the side-lock holder.

Fig. 4.18. A side lock holder is so called because the cutter is held in place with a screw into the side of the holder. It is seen here with a 10mm diameter edge finder.

EXTENDING CUTTERS

Almost any parallel shank cutter can be extended by making an extension shank for it. The extension is simply a round bar with a hole drilled and reamed (or bored) into it to take the shank of the cutter that requires extending. This method is particularly useful for extending FC3 cutters which have very short shanks and equally short cutting flutes.

Make the extension by drilling and reaming the hole and turning the shank diameter at the same time to ensure concentricity. You can fix the cutter in with a grub screw, by soft or silver soldering or even by using Loctite engineering adhesives.

Make sure any grub screw does not protrude from the shank diameter or you could catch your clothes or hand on it. I have seen someone have his overall, shirt and vest ripped off when he caught them on a long grub screw. He spent two weeks over Christmas in hospital all because he lost a short grub screw and replaced it with a longer one.

Autolock holders

Another type of cutter holder is the Autolock holder made by Clarkson. This holder is made for cutters that have a threaded end and parallel shank with a centre in the end of the shank. The pitch of the thread is a slightly undersize 20 TPI Whitworth form and is constant over the entire range including the metric cutters; only the shank diameter varies.

These threaded cutters fit into a special collet which in turn is fitted to a special holder. The holder has an internal back centre to locate in the centre of the cutter shank. The front of the holder has a special nose angle that locates the front of the collet in the milling chuck locking nut.

FITTING A CUTTER TO AN AUTOLOCK HOLDER

Screw the cutter into the collet, fit the collet inside the nose piece and screw the nose piece onto the main body. Adjust the cutter and collet until the nosepiece closes up completely. Undo the nosepiece about ⅓ of a turn and tighten the cutter by turning until the tool shank is firmly in the centre fitted to the holder. Tighten the nose piece the last third of a turn and the cutter will be locked.

I do not recommend using ¼in (6mm) diameter shanks in an Autolock holder, as the small diameter shank is liable to break at the centred end especially when the holder's centre is worn or damaged.

If you are offered an Autolock holder with a damaged centre, I would not buy it. It is almost impossible to remove the centre for regrinding or replacement. It can be done but it takes a very heavy duty press and a lot of brute force. It took me a couple of hours to remove a centre and it finally came out like a bullet. It is far better to buy an Autolock holder with the centre in a good condition.

ER collet holders

As mentioned earlier, the ER holder comes in various sizes. The smallest is ER11 which takes collets from 0.5mm to 7mm. The largest collet size is the ER40 range which goes from 2mm to 30mm.

The ER collet is a double angle collet which can be closed down by up to ½mm (sometimes as much as 1mm) to tighten on smaller shanks than the nominal size. The ER collets are designed to take a plain shank cutter but are fine with cutters that have a flat ground on them.

You can also get small ER holders with

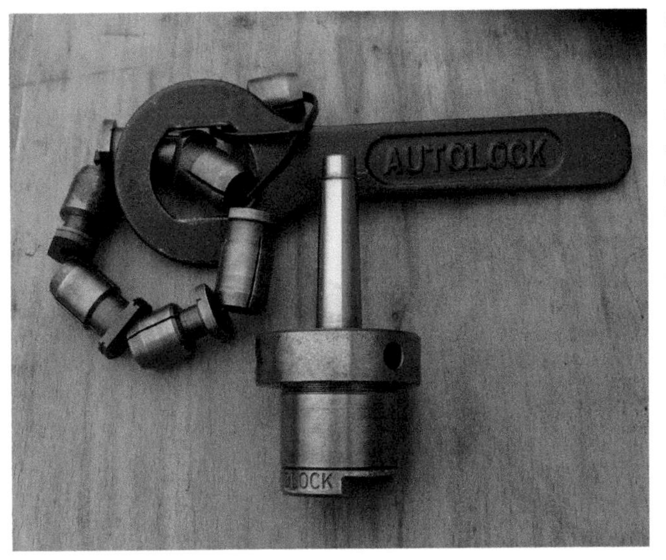

Fig. 4.19. This is a Clarkson Autolock chuck with a set of imperial and metric collets and an Autolock spanner. The collets with the plain ends are imperial collets, and the collets with a groove around the collar (bottom left being an example) are metric collets. I believe the threaded ring can be tightened against the end of the spindle to make the whole assembly more rigid although I have never tried this.

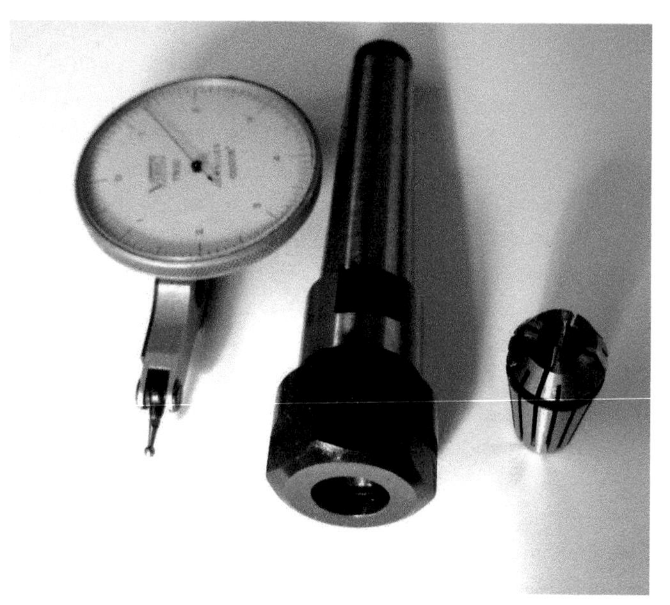

Fig. 4.20. This is an ER collet holder on a 2 Morse taper shank. 3 Morse taper and R8 shanks are available.

Fig. 4.21. This is a selection of ER collets. The groove around the collet needs careful insertion into the collet nosepiece to avoid damage. Tilt the collet as you insert it so the groove enters the nose piece properly. A little careful practice without forcing the collet will soon have you inserting and removing collets easily.

parallel shanks. These are ideal for using as extended cutter holders.

Most small milling machines will have an ER16 (0.5–10mm), an ER 20 (0.5–13mm) or an ER 25 (0.5–16mm) collet set, most often on a 2 or 3 Morse taper. Some of the larger home milling machines may have an ER collet holder mounted on an R8 shank.

Other collets

There are other types of collet holders that are similar to Autolock holders: the Clare system, the Osborne Titanic chuck and one made by Kenametal. None of these systems is as popular as the Autolock system. Cutters are fitted to these systems on the same principle as an Autolock holder; if in doubt, refer to the manufacturer's literature.

FEEDS AND SPEEDS

The correct speeds for different materials

Cutting speeds for milling cutters have been determined through trial and error in industry over many years. They will give an optimum tool life and a reasonable surface finish. Speeds are normally recommended in Surface Speed in Feet per Minute. Speeds and feeds used on an industrial milling machine, where most of the cutters will be made of tungsten carbide, will be a lot

Material, Cutter or Drill Diameter In millimetres	Carbon Steel Stainless Steel Alloy Steels	Cast Iron	Phosphor Bronze Gunmetal	Mild Steel	Copper Hard Brass	Aluminium Soft Brass Nickel Silver
	15 Metres Per Minute 50 Feet Per Minute	18 Metres Per Minute 60 Feet Per Minute	24 Metres Per Minute 80 Feet Per Minute	30 Metres Per Minute 100 Feet per Minute	45 Metres Per Minute 150 Feet Per Minute	60 Metres Per Minute 200 Feet Per Minute
75	64	76	102	127	191	255
50	96	115	153	191	287	382
25	191	229	306	382	1461	764
19	251	302	402	503	754	1006
16	299	358	478	1902	896	1194
12	398	478	637	796	1194	1592
10	478	573	764	955	1433	1911
8	597	717	955	1194	1791	2389
6	796	955	1274	1592	2389	3185

Table 2. Cutting speeds for various materials.

higher than those used where HSS tools are normally used.

Table 2 gives an indication of the speeds that should be used for various common types of material likely to be found in the workshop. These speeds and feeds will vary depending on the rigidity of the milling machine, the cutter and the work-holding setup. This table is suitable for milling and drilling. For reaming use about 25 per cent of the recommended milling speeds.

Table 2 is a guide only; the chances are, the speeds available on the milling machine will not be exactly the same as required. It is a case of 'near enough is good enough'. Set the milling machine to the next available speed lower than the speed shown in the table for the diameter of the milling cutter and the type of material being cut for HSS tools.

For using tipped tools in the small workshop, rather than an industrial environment, I suggest doubling the speed used for HSS tools and use a feed rate of about 0.001in (0.025mm) per cutting tooth as a starting point. So, for a four-flute end mill you would use a feed of 4 thou (0.1mm) per cutter revolution. You can always adjust the feeds and speeds up and down to suit the cutting circumstances.

CUTTING LUBRICANTS

To get cutting tools to cut easily it is advisable to use a cutting lubricant. Cutting lubricant will also extend the milling cutter's working life.

For steel I usually use neat cutting oil as a lubricant, although it does not cool the work as well as soluble oil. Water-soluble coolant is normally used in industry. This is usually oil with a chemical (a detergent) that makes the oil emulsify in water so that it will lubricate and cool the work. Synthetic soluble oils can sting your hands if you have any open wounds. Soluble oils on the milling machine can wash the lubricating oil off the slideways and encourage rust, so use it at your own risk.

Oil can be used for cutting steels, mild steels, stainless steels and high carbon steels like silver steel. For cutting aluminium, the best lubricant is paraffin. It prevents the aluminium from sticking to the milling cutter.

Brass and cast iron are best cut without lubricant unless you need to wash a lot of swarf away. When milling cast iron, if the milling machine is fitted with a soluble oil tank, the cast iron dust will have an affinity for the soluble oil, and if the milling machine is then used on aluminium the free carbon in the cast iron can contaminate the aluminium with black streaks.

I normally apply cutting oil by hand with a small brush using the oil in a cat or dog feeding bowl.

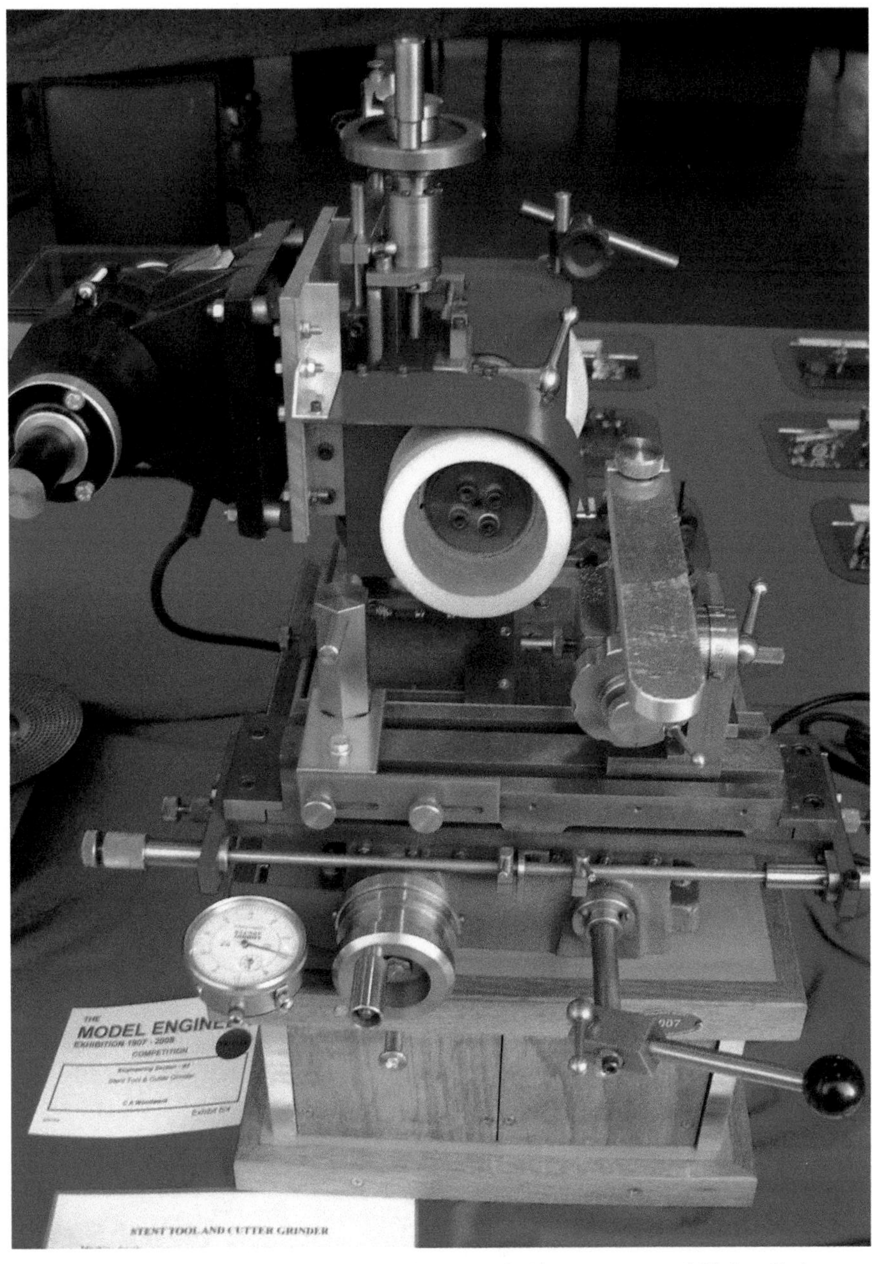

Fig. 4.22. This is a superb example of a Stent tool and cutter grinder. The castings are available from Blackgates Engineering.

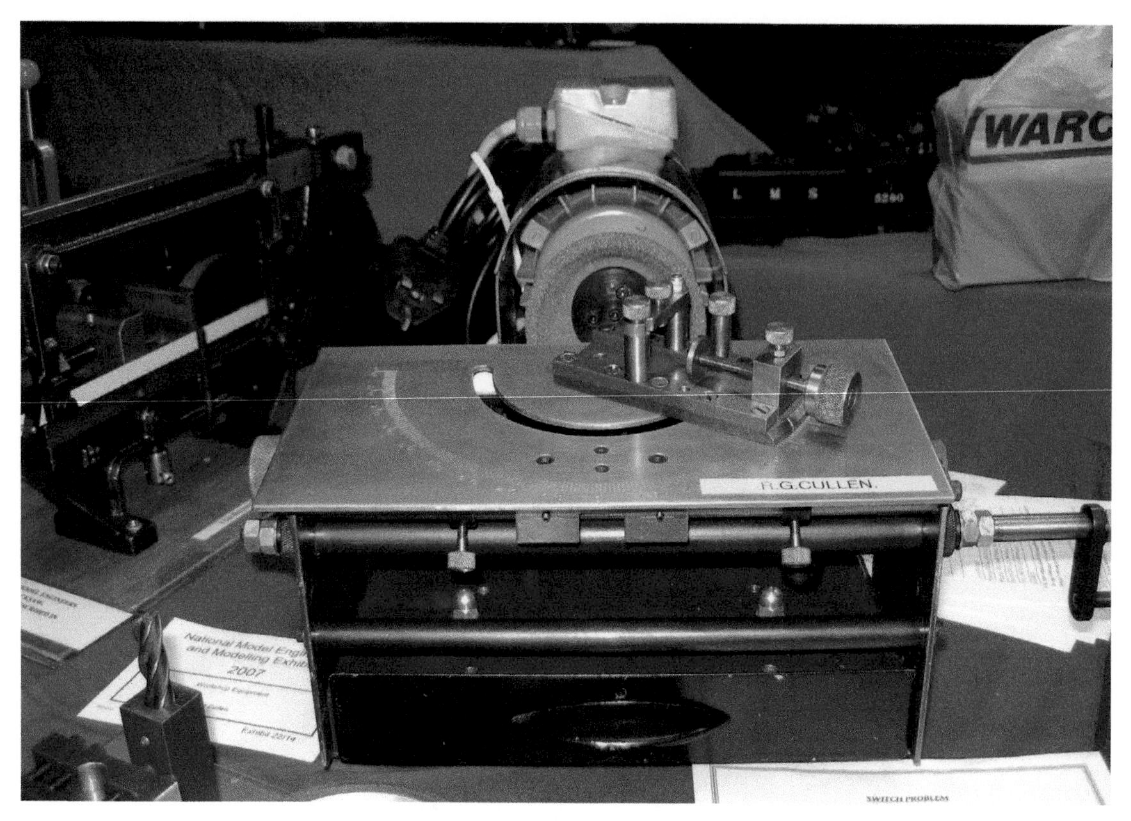

Fig. 4.23. This Worden tool and cutter grinder is mainly made out of sheet metal. It is available from Hemingway kits.

Fig. 4.24. This cutter grinder, designed by Professor Chaddock, is a Quorn. Castings are still available for this grinder.

Fig. 4.25. The drawings for this Collier Caseley cutter grinder are available at a nominal cost from the Bristol Society of Model and Experimental Engineers.

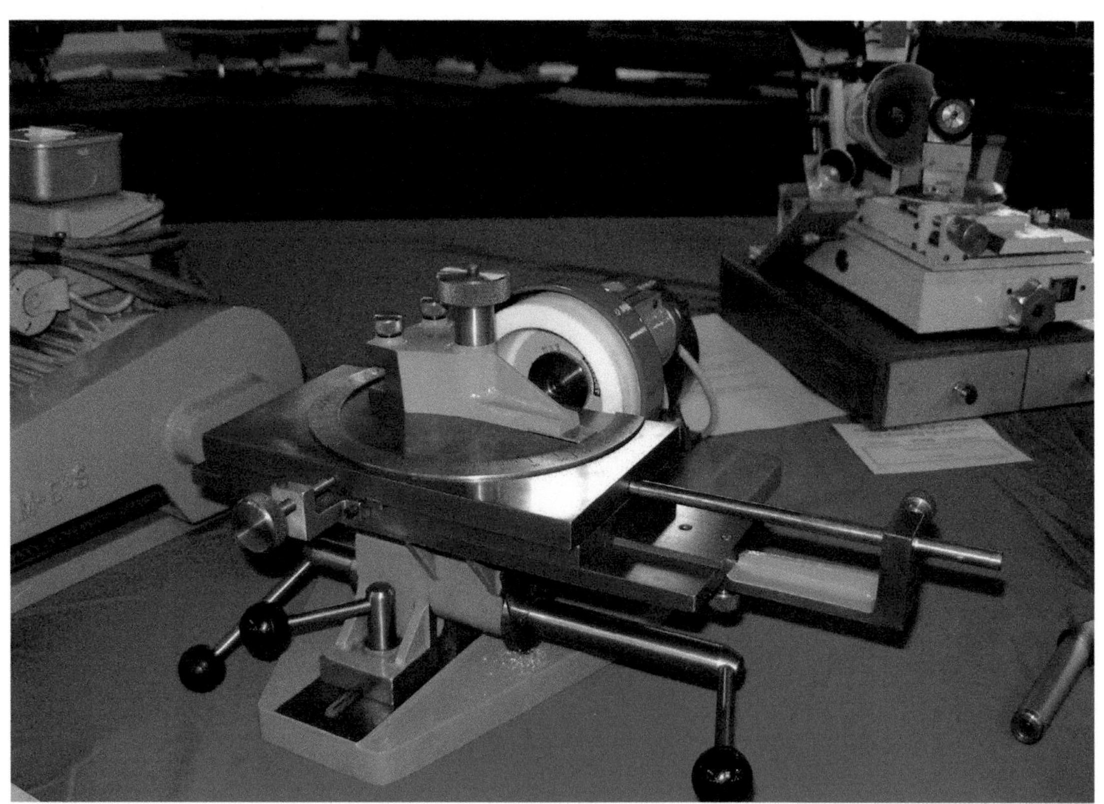

Fig. 4.26. This is a Kennet tool and cutter grinder; castings are still available.

Fig. 4.27. This Tinker tool and cutter grinder is used with an off-hand bench grinder. The kit comprises three substantial castings. Although no longer available, the castings and drawings turn up on eBay occasionally. There are designs on the internet for a smaller version made from stock steel sections.

5 Setting Work True and Finding the Datum Position

The milling machine is the ideal machine for working accurately straight onto metal with the minimum or no marking out. While the machine's dials will allow you to work to a reasonable accuracy, I highly recommend that you fit some sort of digital readout to at least the X and Y axes of your milling machine. These will not only increase accuracy, they should also help you to eliminate errors, such as where you wind the handwheel on an extra turn.

DIGITAL READOUTS

Readouts are quite easy to fit (*see* Chapter 2) and can be purchased at a reasonable cost. It is not quite as important to have a readout on the Z axis where a simple up and down movement is all that is required. It is however quite easy to fit a cheap vernier type scale to the Z axis if you wish.

A readout gives you an exact reading of the position of a particular axis and is unaffected by backlash in the machine slides. It can be set to zero at any point on the machine's travel. Instant conversion from imperial to metric is normally available at the touch of a button.

Absolute or incremental mode

Most digital readouts can work in absolute or incremental mode. Absolute mode means that every position is taken from one absolute point, usually where you set the initial datum point. Incremental mode means you can move to the first position and set the readout to zero, then move to the next position incrementally. Set the readout to zero again and then move incrementally to the next position and so on.

A useful technique I have used in industry is a combination of absolute and incremental. Although this was done more often on a CNC mill, the technique is easily applicable to manual milling machines. A typical use for this technique is to drill holes in multiple components on a fixture. Move to the absolute position and set zero. Set incremental zero. Drill the holes in the first component using incremental co-ordinates. Change back to absolute and move to the next component. Change back to incremental and drill the second component, back to absolute and move to the next component and drill using incremental and so on until all the components are done. This is a very quick method of drilling multiple components accurately.

Other functions

Some readouts have other functions built in, such as a trigonometric calculator, and they can show you the co-ordinates for setting out holes on a pitch circle diameter (PCD) for things such as bolt holes. They can show the co-ordinates of an oblique line of holes on an angle and also show the co-ordinates for contour milling an arc or radius. Other useful functions are a 99-position memory for production work and a centre-finding position where you can divide a measurement by two. This is actually the specification of a simple bottom of the range digital readout so, as you can see, a digital readout is a very useful piece of equipment to have in the home workshop. I would not be without mine.

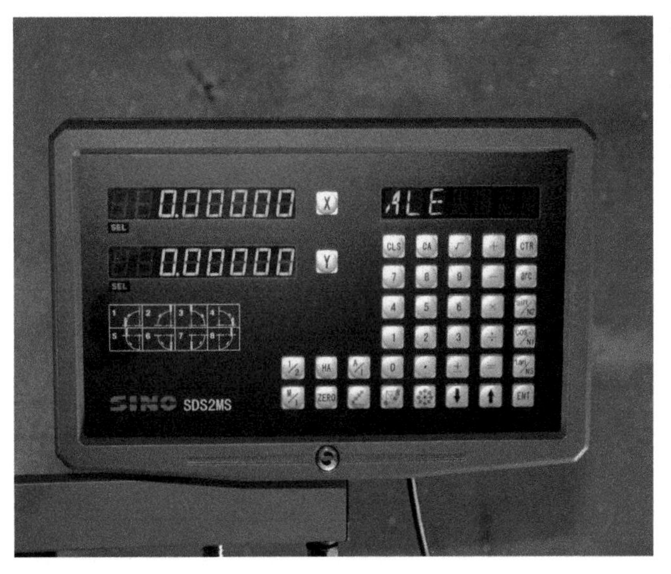

Fig. 5.1. Some sort of digital readout is essential to accurate working on the milling machine.

DIAL TEST INDICATORS

Another indispensable piece of equipment is a dial test indicator. This is used for testing that various milling machine accessories are set up square and parallel to the machine's table, as well as setting up and checking the actual workpieces.

There are two main types of dial test indicator: the lever type and the plunger type dial test indicator, which is sometimes known as an elephant's foot indicator as the working end reminds you of an elephant's foot.

Fig. 5.3. The lever dial test indicator.

The lever dial indicator usually works both ways; it will indicate a forwards or backwards movement. Some lever dial indicators indicate movement both ways while some require you to flick a little lever to change the direction of indication. This type of indicator has a relatively small movement, often a maximum of 15 thousandth of an inch (0.4mm) either way. This is not a problem as you can normally move the stylus to where it needs to be, either by offsetting the dial test indicator in relation to the zero position or by moving the stylus, which is usually held in place by friction.

The plunger dial test indicator has a plunger that slides in and out and is geared internally to a needle that rotates within a dial. The movement of a plunger dial test indicator is usually about ½in (12mm) but it can vary right up to 2in (50mm).

Fig. 5.2. The plunger dial test indicator.

While the lever dial indicator can be used equally effectively for inside holes or outside surfaces, the plunger dial indicator can only be used for external surfaces, although you can get an accessory that fixes onto the non-moving part of the plunger housing that will enable you to clock internal bores or surfaces at right angles to the plunger. The plunger dial test indicator has many uses in the home workshop, but if you can only afford one dial test indicator I would recommend the lever dial test indicator. It will do all you require.

While the plunger dial test indicator will actually read the movement that it is moved, the lever dial indicator does not move the amount it actually says. Because of the lever effect, the stylus moves less the greater the angle it is set on, due to having a spherical end. This is called the cosine error. For this reason the lever test indicator should be used as a comparator where each movement of the indicator should be at the zero point on the dial. One dial test indicator maker (Verdict) has patented a pear-shaped stylus that eliminates this cosine error at angles of up to 36 degrees. However, the easiest way to eliminate most of this error is to ensure the stylus is set parallel to the body, not at an angle.

Setting to zero

Most dial test indicators can be set to read zero by simply turning the external bezel that is normally fitted to the outside of the dial test indicator. The plunger dial test indicator is also available as a digital version and this can be set to zero at any point in its travel by pressing a button.

> **LOOKING AFTER YOUR INDICATOR**
>
> **Lever dial test indicator**
>
> ***There is little in the way of maintenance you can do to a lever dial test indicator other than ensuring that the stylus is tight in its seating and giving it a wipe over with a bit of tissue paper or similar.***
>
> **Plunger dial test indicator**
>
> ***The plunger dial test indicator can usually be cleaned by squirting cigarette lighter petrol down the plunger hole and sliding the plunger up and down a few times. Any remaining petrol will evaporate very quickly.***
>
> ***Do not do this near a naked flame and do not smoke while doing this.***
>
> ***Do not use lighter fuel on a digital plunger dial test indicator.***

Mounting the indicator

There are various ways of mounting a dial test indicator on the milling machine. The

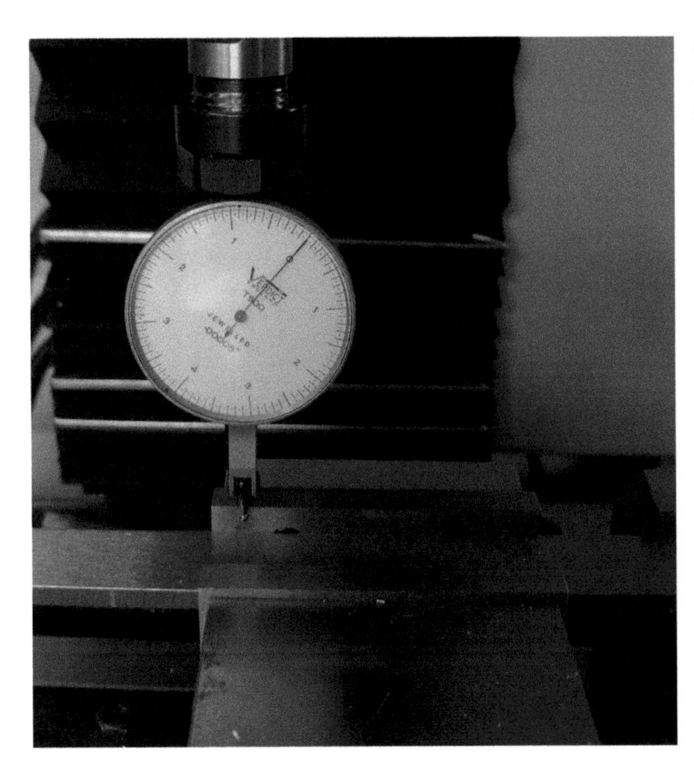

Fig. 5.4. The lever dial test indicator in use clocking up a vice.

most often used method is to hold it in a drill chuck or collet in the milling machine – this is most often used to clock holes true; or it can be placed on a magnetic base to clock faces of vice jaws or angle plates true.

EDGE FINDERS

There are several types of edge finder available. All should be run at about 500–600rpm, except the digital/electronic version which is usually used as a static edge finder.

Wiggler

For many years, the most common edge finder was the wobbler or wiggler type. (This was so called as after it was used to find the edge of a component, it rolled along the edge of the work and wandered off.) The wiggler usually came as a kit with several different ends. The main body accepted the various ends which fitted into a ball-shaped depression in the body, which could usually be adjusted for a friction fit. The usual accessories were a ball, a parallel cylinder, a point for centre finding and an adaptor to take a lever dial test indicator. The ball is for finding the edge of a square piece of work, while the cylinder is used for finding the edge of a round bar. The point is used to find the intersection of two lines scribed onto the workpiece surface. This type of edge finder works well; I used one for many years, but eventually I moved on to the edge-finder or centre-finder type.

Fig. 5.5. The wiggler type of edge finder has been in use for many years. It has largely been superseded by the edge finder type.

Using the edge finder

The edge finder consists of a parallel shank with a cylindrical or top hat shaped end that is held onto the shank with a spring. While the shank is running, bring the edge finder towards the job so the end starts to run true the closer it comes to the edge of the work. When it reaches the workpiece's edge, the end will roll along the edge of the work indicating that the edge has been found. The edge finder end must be square to the work or it will not give a true indication. This type of edge finder is also suitable for finding the edge of a round bar.

Fig. 5.6. The edge finder is used to find the edges of work.

Fig. 5.7. The centre finder is similar to an edge finder but has a point, making it ideal for finding the centre of narrow slots.

Using a centre finder

The centre finder looks like an edge finder with a point. No one ever explained to me how to use this type of edge finder so I have figured it out for myself. Some edge finders are double ended, with a point on one end and a cylinder on the other. I use the centre finder in three different ways. Firstly, you can use it to find the intersection of a pair of scribed lines; secondly, you can use it to find the centre of a slot that is too narrow to get a standard edge finder in by using the centre finder halfway down the point; and finally you can use it to find the centre of a hole in the work.

To set zero using an edge finder, you need to know half the diameter of the end of the edge finder or wobbler. When you bring the edge finder up to the edge of the work and it rolls away, the edge finder is exactly half the diameter of the end away from the edge. Say the edge finder has a 6mm end diameter. This means you have to raise the edge finder above the work and move the edge finder 3mm closer to the edge. This will put the centreline of the edge finder exactly over the centre of the edge of the job.

The digital/electronic edge finder

(This is used without power rotation.) This type of edge finder usually consists of a steel tube with a battery inside and a cylindrical end. You bring the edge finder up against the edge of the work so the end touches the work and completes an electric circuit through the work, the vice, the frame of the machine, through the spindle into the top of the edge finder. Because of this need for conductivity, the digital/electronic edge finder is not suitable for non-metallic objects such as plastics.

TRAMMING THE MILL

Tramming the milling machine refers to setting the machine's head square to the machine's table. This is most often done with a dial test indicator in the machine's chuck. The dial test indicator is set to a reasonable diameter of swing, say, about 4–6in (100–150mm) with the stylus clocking up the top of the table. To avoid the test indicator bouncing up and down over the T-slots, a parallel should be placed on the table and the test indicator should be set to the top of the parallel.

When the dial test indicator is swung at equal 90 degree positions, the amount the head is out of square can be determined and the head adjusted until the dial test indicator reading is zero at all four positions. This assumes the head can be rotated both ways. If it cannot be rotated both ways and the head is out in the direction the head cannot be rotated, then you will have to square the head up using shims or some other method until the head is square. Once this is done, the head should remain square in the direction that the head cannot be moved, the shims keeping the head square in this direction.

Squaring the head up in the moveable

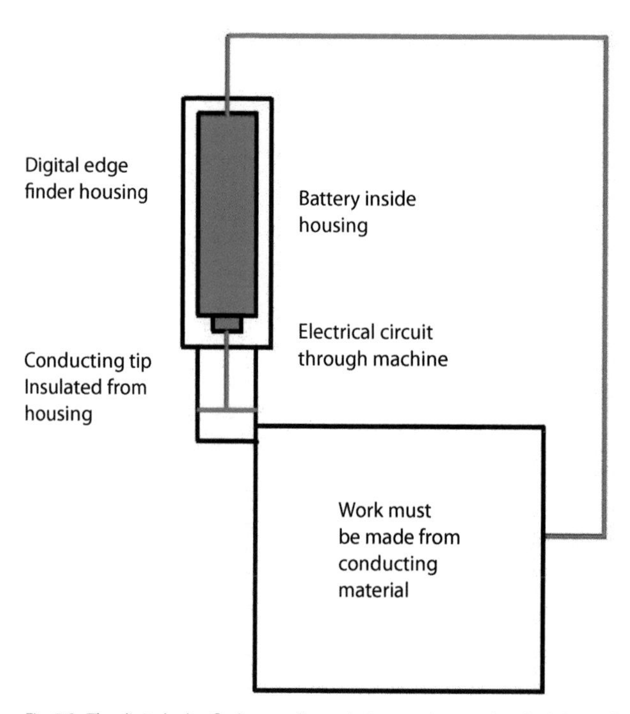

Fig. 5.8 The digital edge finder usually works by creating an electrical circuit through the work, the machine and the spindle. For this reason it is not usually used on non-electrical conducting components.

Fig. 5.9. *You should check the tramming of the milling machine's head before doing an accurate piece of work and after doing any heavy metal removal.*

Fig. 5.11. *The threaded rod in the middle is the milling machine's down feed stop. The threaded collar can be wound up and down to set the drill's depth.*

direction is usually a matter of trial and error, moving the head an absolute minimum until it is square. Once it is square, if at all possible do not move the head out of square. This may not always be possible but try to do this. Eventually the head will become out of square, possibly because of a heavy cut, so do check regularly that the head has not moved, especially before a job requiring extreme accuracy.

X, Y, AND Z CO-ORDINATE SETTINGS

A milling machine's co-ordinate system is designated in three axes, each of which can be sub-designated + or -. So, you can have X+ and X-, Y+ and Y- and Z+ and Z-. The + part of the system is not normally used in practice because X+1 is the same as X1, but X-1 is not the same as X1, and similarly for Y and Z.

Looking from the front of a milling machine, X- is the bit of table to the left of the spindle; X+ is the bit of table to the right of the spindle. Y- is the bit of the table nearest to you in front of the table, Y+ is the bit behind the spindle. Z+ is above the bottom of the cutter and Z- is below the bottom of the cutter. These are the six directions of movement of a milling machine.

These co-ordinates are very useful when working with a digital readout. You can set the readout to the zero position that you require and then move in the + or – direction in any axis and get a direct readout of the position you are at.

If you do not have a readout on the Z axis, you could set the Z depth of the cutter in various ways:

1) Wind up the table (if a knee machine) so the cutter just touches the work. Then wind up the table by the amount the cutter needs to go into the work.

2) Place a spacer (possibly a drill shank) between the quill travel stop and touch the cutter onto the top of the job. Remove the spacer, and the cutter is set to the correct depth.

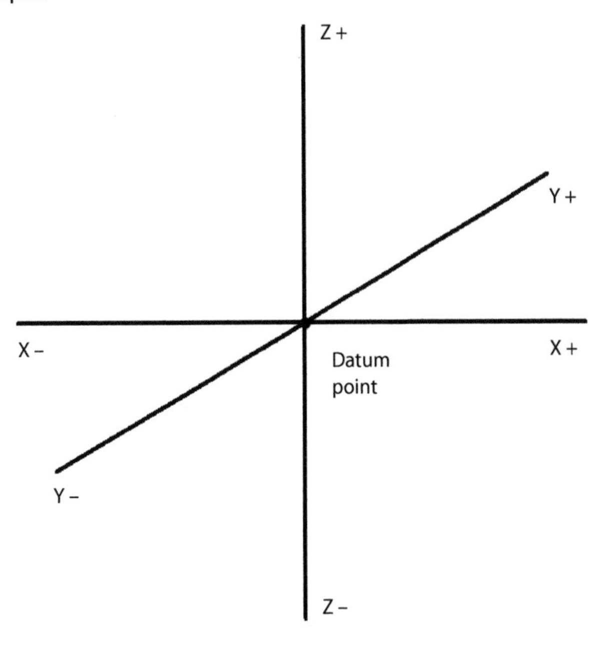

Fig. 5.10. *The six directions of movement of a milling machine are shown here.*

3) Set the Z by raising the table against a plunger dial test indicator mounted on a magnet.

Edge finding from the vice jaw and parallels

Using an edge finder or wobbler, it is very easy to find the edge or centre of a workpiece. However, you cannot always find

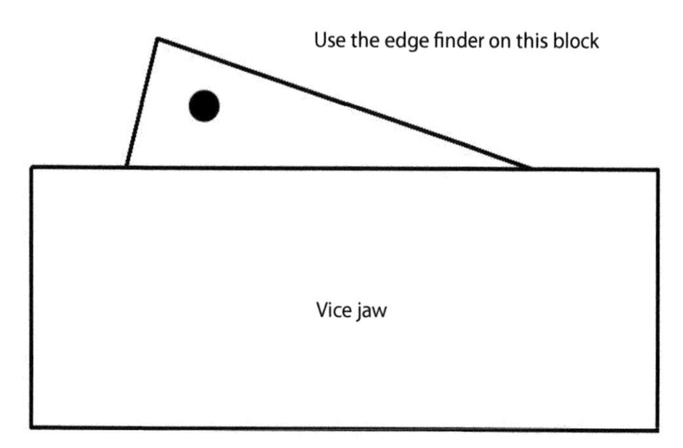

Fig. 5.12 Using a piece of bar on an angle so it is above the vice jaw will enable you to find the exact edge.

the edge of a job, for instance because the job is lower than the vice jaws. In this case, there are two ways to find the edge:

One is to temporarily put the workpiece in the vice higher than normal. The workpiece does not have to be flat in the vice; as long as it presents a flat face to the wobbler it can even be held on an angle in the vice.

Another method is to put a parallel into the vice and use the edge finder on that. Either method will do what we require.

SETTING THE DATUM OF A WORKPIECE

Setting from a horizontal hole

If we have to set a horizontal hole central put a bar through the hole and wobble off the bar to find the centre of the hole. This is very useful when we have to drill mounting holes in symmetrical objects like bearing blocks.

Fig. 5.13. If you wobble off the bar through the main bearing hole you can find its centre.

Toolmaker's buttons

Another method of setting the datum of a workpiece is by using toolmaker's buttons. Toolmaker's buttons are readily available on places like eBay, but they are also easy to make.

They are literally a turned tube counterbored to take a screw to screw them to the workpiece. You set the buttons using micrometers and/or spacers so they are the correct distance apart or from one edge of the workpiece. Then you put the workpiece into the milling machine vice and clock up the outside of the button with a lever dial test indicator. Then you can proceed to drill, bore or whatever at the position of the toolmaker's button, and then move to the next button and so on.

Hollow toolmaker's buttons are similar to normal toolmaker's buttons, but they are longer and the internal counterbore is dead true to the outside of the button. These are useful when the buttons are placed

Fig. 5.14. Toolmaker's buttons are screwed to the workpiece in the exact position where the hole is required so you can clock them up true.

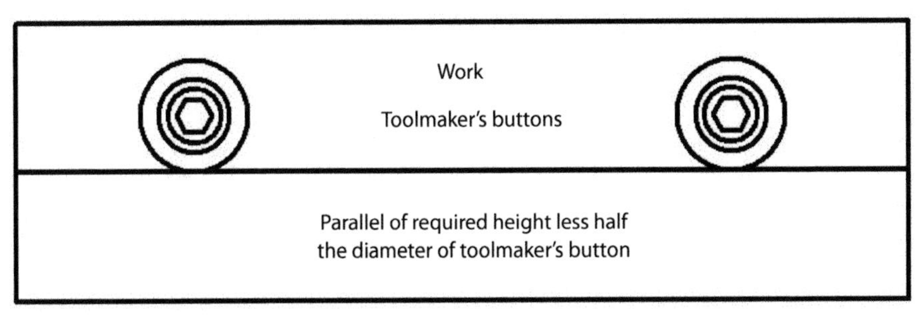

Fig. 5.15 Toolmaker's buttons can be set on top of a parallel to ensure they are set at an exact distance from an edge.

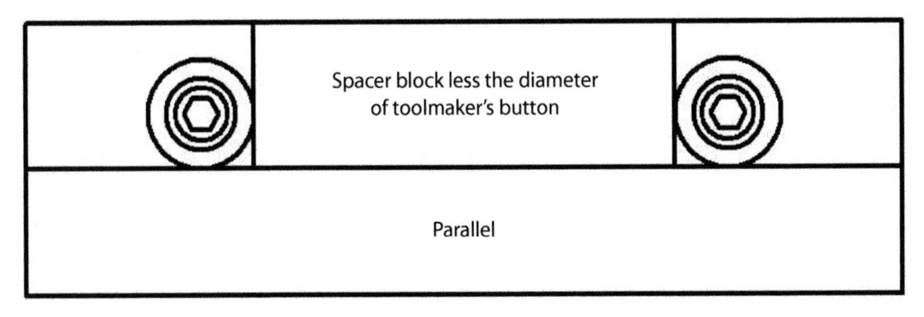

Fig. 5.16 Toolmaker's buttons can be set with a spacer between them to ensure they are set at an exact distance apart. This can be combined with the parallel method.

too close together to allow a clock stylus between buttons, as they allow you to clock the inside diameter of the button.

SETTING THE VICE AT AN ANGLE

Put a protractor set to the required angle in the machine vice. Run a test indicator along the angular protractor blade until the clock reads zero all the way along and then clamp the vice down. The vice will now hold the work at the right angle.

Fig. 5.17. A simple sheet metal protractor can be used to set work on an angle.

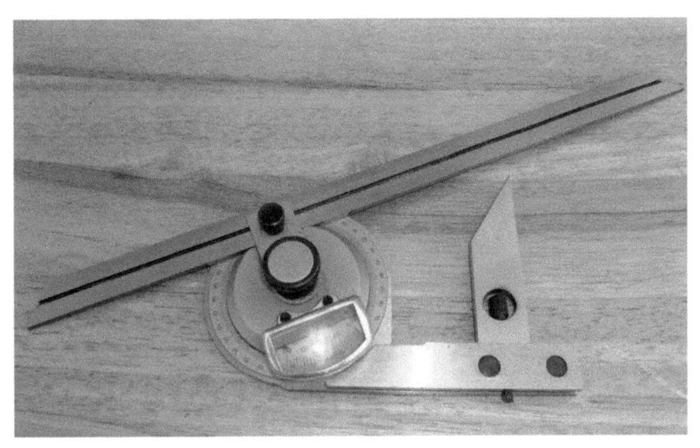

Fig. 5.18. A vernier protractor is a more accurate way to set work on an angle.

Sine bar

A sine bar is useful for clocking a vice on an angle. Simply put the sine bar into the vice with the required parallel spacer between one end of the sine bar and the vice jaw. You will probably have to put a bit of round bar in the vice to nip the sine bar in the correct position. Then clock along the sine bar and clamp the vice down on an angle.

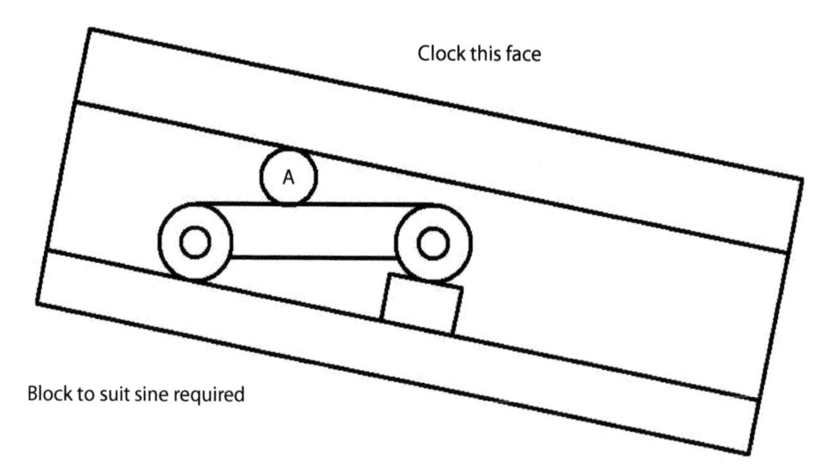

Fig. 5.19 The sine bar is being used to set the vice on an angle. Bar A is just to hold the sine bar against the vice and packing block.

Digital angle gauges

Another method of setting work vertically on an angle is by using a digital angle gauge. This is a small 2in (50mm) square electronic gauge that can be set to zero and then turned to any angle. Put the component into the vice and zero the digital angle gauge on top of it. Now you can rotate the component on an angle in the vice using the digital gauge to ensure the angle is correct.

Fig. 5.20. A digital angle gauge can be zeroed at any position and then used to set the job to any angle.

6 Work Holding in the Vice

EQUIPMENT

Machine vice

The main vice you will be using will be the machine vice. This is normally a substantial item made from iron castings. It is worth buying a good quality machine vice as it should last a lifetime.

The jaws of a machine vice should preferably be hardened and ground, although soft jaws are very useful as well. The jaws must also be vertically aligned with the milling machine's table and head. This can be checked with a lever dial test indicator. If you have a knee machine you should check the verticality of the vice by winding the table up and down and also by moving the quill or head up and down. Both should show the same (zero) reading.

Normally you would put the vice onto the milling machine either square or at 90 degrees to the travel of the X axis of the mill. If you get into the habit of clocking the vice up square you are less likely to make a mistake. If squareness of the job does not matter, still clock the vice up, as the next job may be important and if you put the vice on roughly square but not clocked, you may forget the vice is not true and scrap the next workpiece. In industry, the next person may come along and assume the vice is square and they may scrap their work and they will not be very pleased. Yes, I know you should not assume the vice is square but in a professional environment you expect professional workmanship.

Fig. 6.1. This Abwood vice is the Rolls-Royce of vices. It is brand new, the discolouration is protective grease. I purchased it from the local free-ads paper. They are not cheap but will last a lifetime.

If you must put the vice on out of square for quickness, either leave a message on the vice or remove the vice after use. It can remain on the milling machine, but remove the clamping bolts and turn the vice at an angle so people can see it is out of square.

When clocking the vice with the lever dial test indicator, concentrate on getting it correct at both ends of the vice jaw. It may be that the middle of the jaw is slightly bowed. If the bow is a matter of a thou or two this will not matter, but any larger bow should be investigated and corrected before moving on.

The majority of the work you will do will need the vice set at 90 degrees to the milling machine's X axis. I do not recommend that you use a swivel base if your vice has one. This raises the vice off the table, it reduces headroom height and it makes the setup a lot less rigid.

If you have two vices the same, set them up on the milling machine side by side. You will find this very useful. Firstly you can hold long workpieces in both vices at the same time. Secondly you can hold a parallel tightly in one vice so it cannot move in the second vice. Thirdly, you can use the left-hand vice to hold a work-stop and the right-hand vice to clamp the work. If the bases of the vices are at slightly different heights, just machine a pair of parallels to suit so the height over the top of the two parallels is matching.

Fig. 6.2. If you have them, two vices will make many operations easy, especially when machining longer components.

Soft vice jaws

Soft jaws for the vice can be made from steel or aluminium. You can use soft vice jaws in several ways. The obvious way is to skim the inside of the jaws once they are in the vice. Just a light cleaning-up cut is all that is required. Place a bit of round bar in the jaws and tighten up before skimming. The round bar will enable the moving jaw to roll up the same as it will do if the bar is not there. You could also mill all the way along the soft jaws creating a channel to hold the work in.

Finally, you could machine a nest in the vice jaws to take a component. A circular nest made with a slot drill or similar cutter will allow you to clamp a circular component in the vice; you could machine a square for square or rectangular components; and

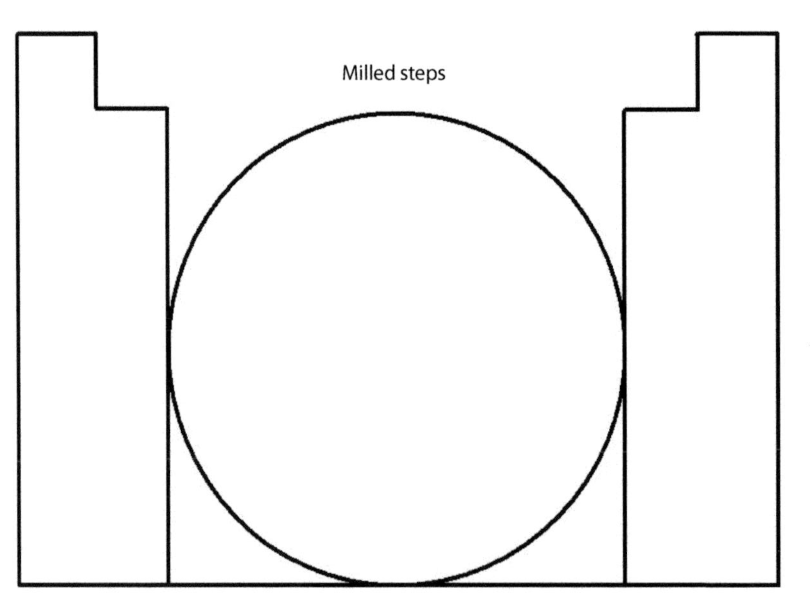

Fig. 6.3 Machining soft vice jaws. The round bar allows the moving jaw to ride up, similar to the way it would behave if there was a workpiece in the vice.

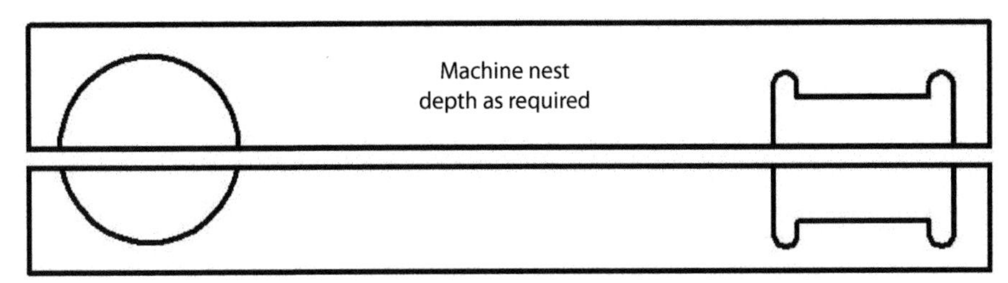

Fig. 6.4 Machining soft vice jaws. The round nest will take bar components while the square one takes square components.

likewise you could machine a nest of any shape to suit any shape of component.

> **A VICE AND A CHUCK**
>
> ***If you put a vice and a chuck (rotary table or dividing head) onto the milling machine table at the same time, with a digital readout you can set datums for two or even three items and switch between them without resetting. This can save a lot of time.***

A workpiece stop

To aid repeatability of loading work into a vice, a workpiece stop is very useful. A stop can be as simple as a magnet on the end of the vice jaw. Some vices will have a drilled and tapped hole, either in the body of the vice or the fixed jaw so a stop can be bolted onto the vice. Another method is to clamp a round bar on top of a parallel between the vice jaws. This does entail drilling and tapping a small hole in the body of the vice, but you might get away with drilling and tapping the jaw if it is soft.

Another method, as mentioned earlier, is to use a second vice with a stop in it. This does not need to be a narrow stop; you can use a stop up to the capacity that the vice will open up to.

Say you want to machine 50 lengths of bar to the same length. Put a wide stop into the second vice, using a bit of sheet metal. Machine the edge of the metal so that it is square to the vice jaw. Now, every bit of bar touching the stop will be at the same position endwise. It is now a simple operation to load as many workpieces as you can and machine the ends square. All will end up at the same length. You can use the same idea but hold the stock of a square in the second vice.

The toolmaker's vice

The toolmaker's vice is usually ground all over and comes in several different sizes. Often these are called pin vices (not to be confused with pin vices that take round stock) because the moving jaw is clamped by tightening up an Allen key which clamps the moving jaw into half a pin hole under the vice. To clamp these toolmaker's vices down, you will probably need to make some clamps with rounded ends that fit into the pin holes, or thin clamps that will fit the clamping slot cut into the ends of the vice.

Home-made long piece table vice

A simple vice that will hold work flat to the table can be simply made from some offcuts of square bar, although rectangular bar would be better. You should make two fixed jaws and a moving jaw. The fixed jaws are just clamped down to the machine table at positions to suit the length of the workpiece, leaving room inside for the workpiece and the moving jaw.

Drill and tap a fixed jaw for two or three clamping bolts parallel and horizontal to the machine table. Insert two or three Allen screws and you can tighten up the moving jaw by the use of these screws. A couple of clamping holes in the moving jaw will enable you to nip the moving jaw down when it is tight. Do not tighten this vice too much or you may distort or even break the table – this is unlikely, but do use a bit of common sense when tightening up this type of vice.

Fig. 6.5. The toolmaker's pin-type vice.

WORKING WITH THE VICE

Squaring up a workpiece

One of the commonest jobs on the milling machine is milling a block of metal square, parallel and flat.

Method 1

If you can, start off using a piece of metal thicker than needed. This way, you can hold the workpiece on the waste piece with the finished block all above the top surface of the vice. Then you can machine across the top of the block and around all four of the edges at one setting. Then deburr the block, turn it over in the vice and machine the top surface so the final face is flat and to size. This way five of the surfaces are guaranteed to be square with each other, and if you are careful the final surface will be square and parallel as well. You can use this oversize block method to machine five out of six faces on any shape of block.

Method 2

Another method is to machine the block a face at a time. Put the block into the vice and machine the largest flat face to clean up. Place this face against the fixed vice with the longest edge showing. Clamp using a round bar between the moving vice and the workpiece. Machine the edge that is showing so it cleans up. We now have two faces square to each other.

Flip the block over, first face still against the fixed jaw, second machined face down in the vice. Tighten the vice up, still using the round bar, and machine the side to finished width. Now we have three machined faces square and parallel. Put the workpiece into the vice, on parallels and machine the fourth face flat, true, parallel and to size.

There are several ways to machine the two remaining end faces. The first is to put the workpiece in the vice with one end hanging

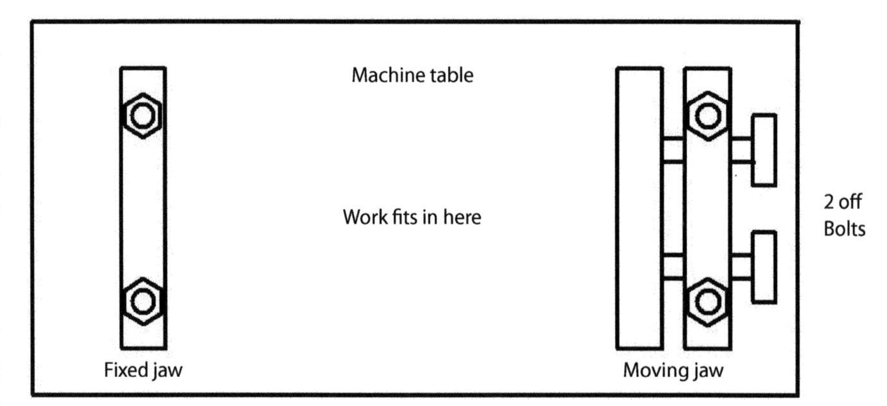

Fig. 6.6 The home-made long piece vice. The fixed jaw, moving jaw and end clamping block are made out of square or rectangular bar. The fixed jaw and clamping block are bolted down to the table while the moving jaw just rests on the table and is fixed when the two end bolts are tightened up.

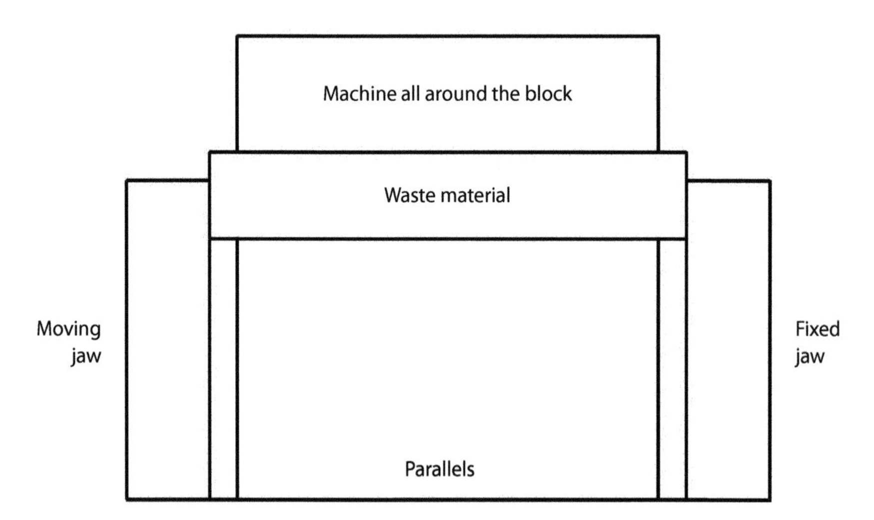

Fig. 6.7 Method of machining five faces of a block square at the same setting. You can machine any profile you like before removing the waste material. If the resulting profile is hard to hold you could hold it in a shaped nest.

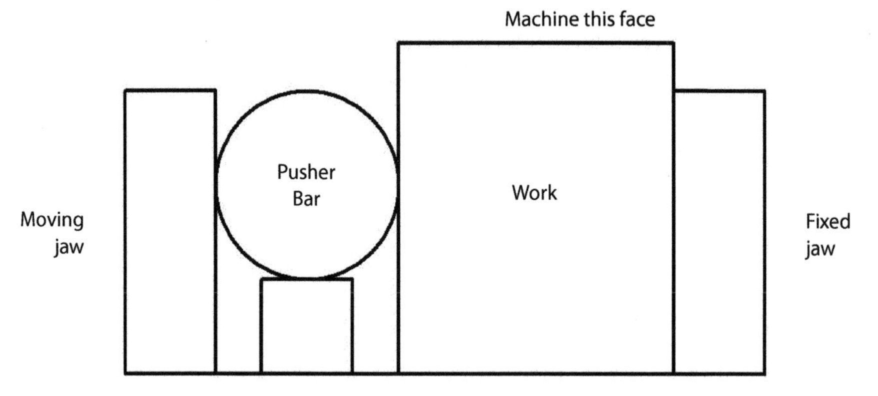

Fig. 6.8 The sequence for cleaning up a square bar: Mill the first face square. Put this machined face against the fixed jaw and machine the next face square. Turn the workpiece over and machine the third face square. Remove the round bar and machine the last face square.

Fig. 6.9. Milling the ends of the bar square: First mill a flat on the end of the block. Then put this flat face down in the vice resting on a small round bar. Then clean up the other end, turn the block over and clean up the other end.

outside the vice. You can now machine the full depth of the end face with an end mill. Turn the workpiece round so the other end of the workpiece is hanging out of the vice and machine to finished length.

Method 3

If the depth of cut is too great for the length of your end mill, proceed as follows. Put the block in the vice and machine part of one end square. You only need to go about ½in (12mm) deep – the depth is not critical. Put the block into the vice with a small round bar under the machined step in the end and tighten up. You can now clean up the remaining unmachined end, which will be square with the block. Turn the block over in the vice (it does not matter if you leave the round bar under the work or not) and clean up the final end square and to size.

Method 4

An alternative method is: once you have machined all four of the faces, put the block vertically in a pair of vee blocks in the vice. As long as the vee blocks will clamp the block vertical without touching each other, you can machine the end square and turn it over to machine the other end. If the workpieces are small enough, you may be able to do them four at a time. If the vee blocks touch each other and the workpiece is not clamped, you can make the workpiece bigger by putting a piece of angle between the vee block and the workpiece.

Milling a round bar square

Often, you will not have a bit of square stock the right size so you will need to mill it from out of the round. This is simple to do: start with a piece of bar ×1.450 (say ×1.5 to be sure) larger than the square bar required. Put the bar in the vice and touch on with the cutter. Mill off 0.275 × the diameter. This is the first flat taken to width. Put the bar into the vice with the flat against the fixed jaw. Remove another 0.275 × the diameter from the top of the radius. The flat should meet up with the first flat. If necessary take off a few more thou. You now have two flats at right angles.

Turn the bar over so it rests on the bottom of the vice or on parallels. The round part of the bar should again be against the moving jaw. Mill off the radius until the bar is the correct thickness. Finally, mill off the last radius. Put the bar into the vice with the radius at the top. Mill to size. You should now have a nice square bar. Finish the ends to length as described for the square bar.

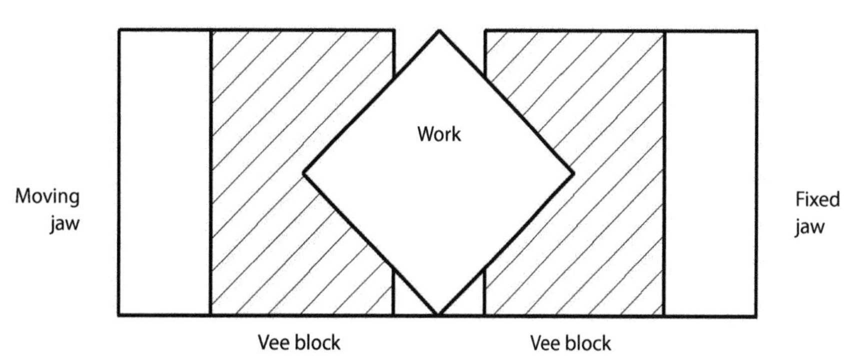

Fig. 6.10 Milling the ends of the bar square in a pair of vee blocks. You could also do the same with only one vee block.

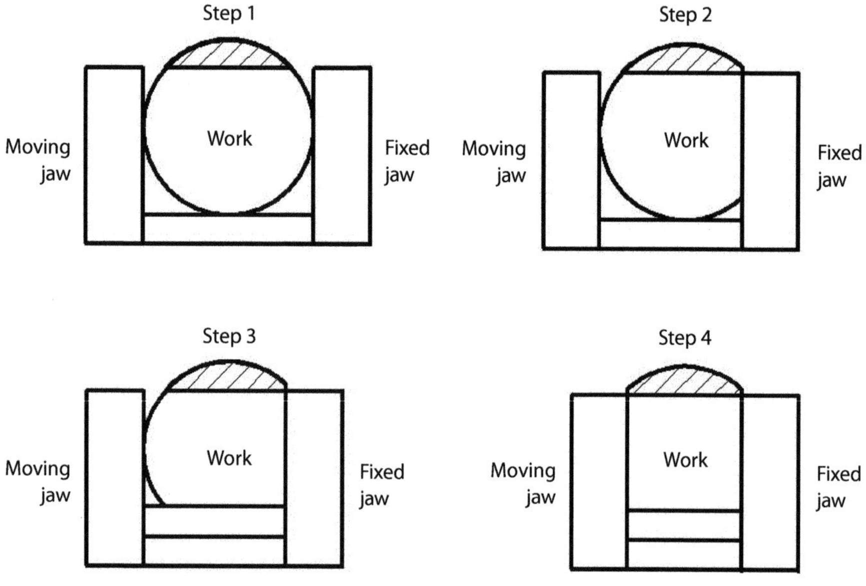

Fig. 6.11 The sequence for milling a round bar square: Mill the top face. Mill the second face, mill the third face and finally mill the fourth face.

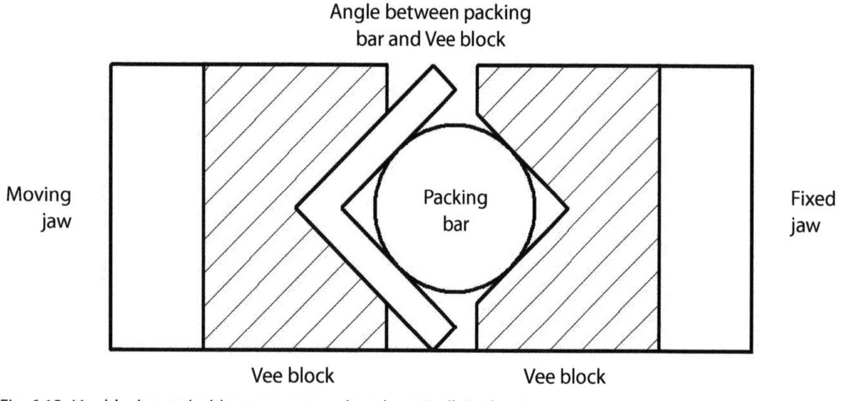

Fig. 6.12 Vee blocks can hold square or round work vertically in the vice.

Using vee blocks to square up work

We touched on this method when we were machining blocks square. You can drop the workpiece into an upright vee block in the vice to mill the ends square. You can drop it into two vee blocks to do the same. You can also mill the ends of a bit of angle by putting it into the vee block and putting a bit of round bar between the V in the angle and the moving jaw. Finally, you could machine the ends of a bit of round bar square to the axis while it is held between two vee blocks or a vee block and a vice jaw as mentioned previously.

Using parallels

Parallels are very useful in the machine vice. They usually come in sets of different heights, often in a plastic box. Each pair will be ground to the exact same height, and the ends are usually left rough as sawn off. The best type of parallel is hardened and ground. If possible, get parallels about ⅛in (3mm) thick as these will support the work while taking up minimal room in the vice. To fix the parallels to the vice jaws (to stop them from falling over) put a smear of grease on the jaws, put a compression spring or two between the parallels or use a bit of foam rubber. Each of these methods will hold the parallels against the fixed and moving jaws.

Place the work on the parallels, up against a stop if necessary, and gently tap it down with a soft hammer, or you could use an ordinary hammer with a bit of brass, copper or aluminium on top of the workpiece to protect any finished faces.

Fig. 6.13. Parallels are usually bought as matched pairs and are often supplied as a set in a fitted or plastic box.

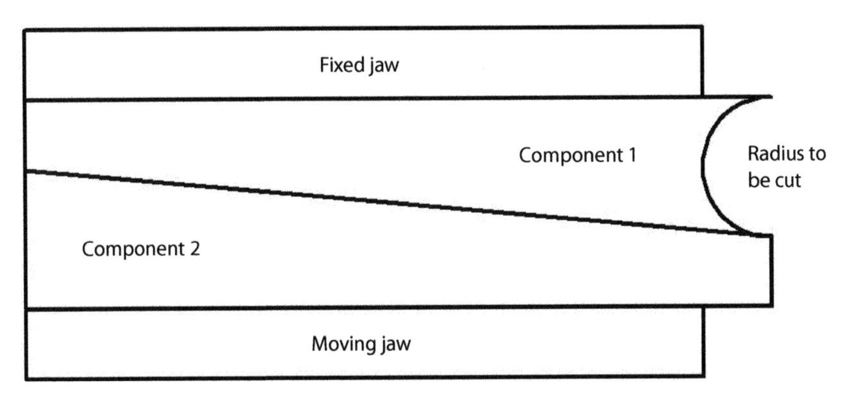

Fig. 6.14 These tapered shapes are handles for a spanner. A standard socket is welded to the radiused end. The tapered handles are placed together so they make a parallel that can be held in the vice to machine the diameter for the socket.

Clamping tapered work back to back

If you have two or more pieces of work with the same shallow taper, you can clamp them back to back in the vice and, as long as the taper is quite shallow, they will be held firmly.

Using 5C collets

There are some widely available collets called 5C, which are draw-in collets designed to be used in a collet holder with a drawbar. They range from 3/64in to 1⅛in (metric collets are also available). You can also buy 5C collets to hold square and hexagon shaped components.

A 5C 'emergency' collet can be obtained, which has no proper hole machined in it, just three rods, one between each slit. In use, you put a rod in each of the slits, tighten up the collet, then you can machine a location in the collet to take the work. The location can be any shape or size within the limitation of the collet. To use the emergency collet, remove the three rods and you can then tighten up the collet as normal.

There are expanding 5C collets where you can tighten a screw to expand the collet to grip internal bores, and there are extra-large collets that can be bored out larger than the 1⅛in standard collet.

Angular machining

You can get vices that tip up on an angle so you can do angular machining without disturbing the alignment of the mill's head. These vices are very useful for milling the edges of dovetail slides, for example.

The main tilting vice usually available is basically a small vice with a pivot at one end and a couple of sheet metal clamps at the other end to support the vice on an angle. The other type has a radius machined on the base to match another radius machined on the vice. This radius enables the vice to

Fig. 6.15. This angle vice has sheet metal stays to support the vice on an angle.

tilt vertically. This type of vice is more rigid than the pivoting type, but it is also more expensive. Either type of tilting vice will do for light work.

DIY ANGLE VICE

To make a poor man's version of an angle vice, hold a small toolmaker's vice on an angle in your main machine vice – it will do the same job as an expensive angle vice. The only thing to remember is to take the vice handle off the main vice and walk across the workshop and leave the handle where you cannot pick it up and undo the main vice and lose your angular setup.

Fig. 6.16. This angle vice is more substantial than the one with sheet metal stays.

7 General Machining

MILLING APPLICATIONS

Using jacks

Jacks are useful to support the work being machined. There are two main types of jack. The first is a solid piece of metal (usually round) with a threaded hole down the middle to take a bolt. The bolt is locked by a nut that tightens against the jack body. The other type is a hollow jack, a body with a plain hole through. An adjustable nut and a bolt allow for adjusting to different heights.

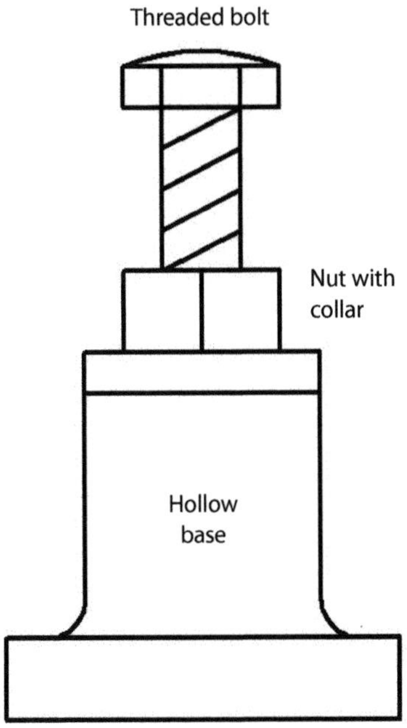

Threaded bolt

Nut with collar

Hollow base

Fig. 7.1 Machine jacks are easy to make in the home workshop. This is an example of a hollow jack. The threaded bolt goes into the hollow base through a clearance hole. The jack height is adjusted by the shouldered nut on top of the base.

In use they are placed under the work that needs supporting and the bolt screwed out to support the work. There are many uses for the jacks; a couple of pairs are very simple to make in the home workshop.

Machining a flat surface

Face mills

Although we have machined a block square (*see* Chapter 6), we did not look at the cutters to do this. You can buy face mills to clean up an entire face in one pass, but they may be too large a load on a small milling machine. By all means try a good carbide-tipped face mill if you can borrow one, but do not buy one until you know if your milling machine has the power and rigidity needed to use it.

End mill or slot drill

A good alternative is a large end mill or slot drill. The largest cutter shank a hobby machine is likely to take will be a 16mm shank, and cutters up to 20mm diameter are readily available with a 16mm shank. If you have an old cutter that may be slightly blunt, a small 1mm × 45 degree chamfer ground on each tooth will leave a good finish on the machined face. This chamfer will also decrease the load per tooth, allowing you to take slightly deeper cuts. Take a cut all over the surface to be machined and then run over it again taking a shallow cut of perhaps a couple of thou. This should give you a good finish.

Flycutter

Another method of generating a flat surface is to use a flycutter, a single-tooth cutter that

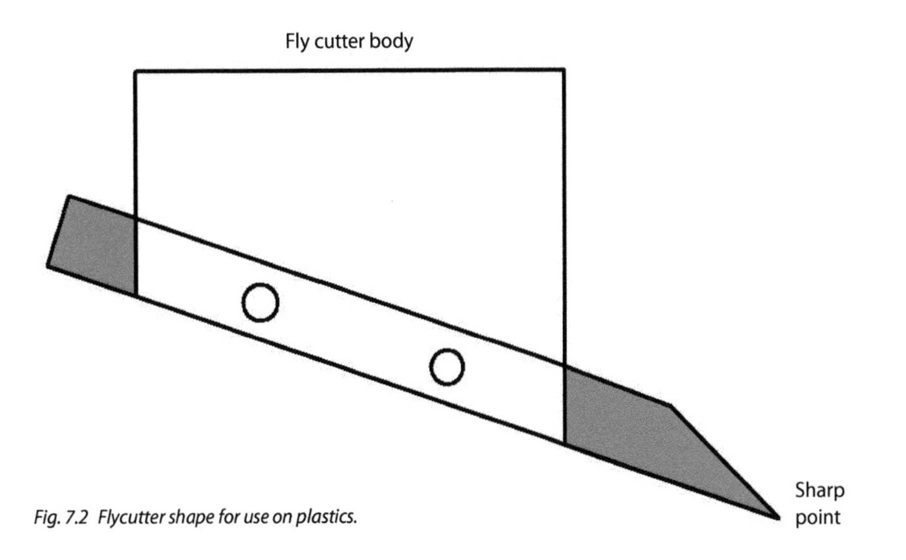

Fig. 7.2 Flycutter shape for use on plastics.

skims right across the surface in one go. Because the flycutter has a single tooth, it is a much lighter load on the milling machine and is much less likely to deflect. If using a flycutter on plastic I recommend a very sharp angle on the flycutter so it slices off the plastic rather than pushing it off.

Machining a slot

The way I usually cut a through-slot is to put an undersized drill through both ends of the slot and then put a size slot drill through at each end. I then mill away the centre of the slot with an undersized slot drill and finally finish both sides of the slot with an undersized slot drill or end mill. This ensures a size slot with correctly rounded ends and no key-holing effect. You can cut slots to depth using the same method, but after machining them run a size cutter all along the slot to ensure the bottom has been cleaned up flat.

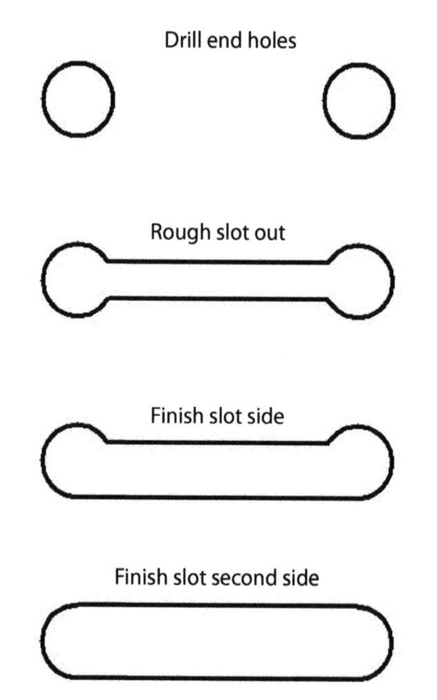

Fig. 7.3 Take the ends of a slot to size before removing the middle.

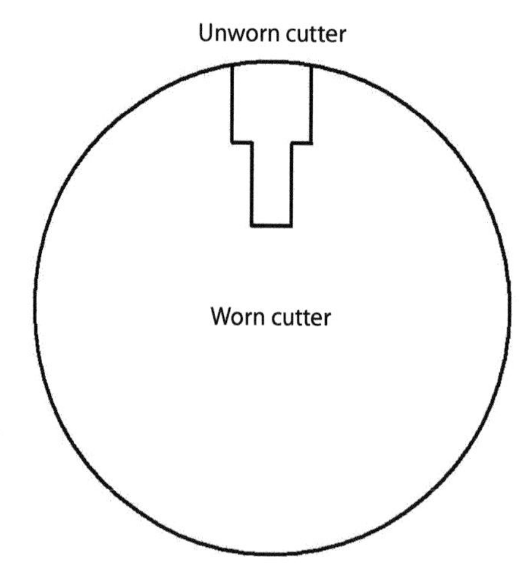

Fig. 7.4 Using a slot drill with a worn end will create a stepped keyway. This diagram is greatly exaggerated. Always use a new cutter for each different depth of keyway.

Machining a keyway

It is easy enough to machine a keyway, but a little care is needed. Firstly, the slot drill should be in good condition. If the slot drill has previously been used make sure it is not part worn on the lower sides of the cutting edge. This could lead to a stepped keyway and cause difficulty in fitting the key. If in doubt, use a new slot drill for each different depth of keyway required.

Slot drills (the two-flute and the FC3 throwaway three-flute ones) are designed to cut about a ½ thou small, which is ideal for a keyway. The keyway should be cut to depth in one pass and in one direction, using a very slow or hand feed.

Machining T-nuts

You can buy T-nuts, but they are a simple exercise in machining so I suggest you make a dozen or so.

First, measure the T-slot and make a cross section view. Cut some lengths of bar sufficient to make four or five T-nuts from each length of bar, leaving sufficient material to hacksaw the individual T-nuts off when they are machined.

Machine the bottom face of the T-nut – this is the bit that ends up at the bottom of the slot. Machine both sides so the bar is the finished width required. Drop the bar sideways onto a parallel with sufficient above the vice jaws to make the side cut-outs. Rough the first side cut-out out and turn the bar over and rough out the second side.

Measure the overall width of the remaining centre of the bar. Take note of the size of the bar and lower the cutter by half of the difference to bring it to size; say the middle of the tee nut needs to be 12mm wide and the bar measures 13.5, the difference is 1.5 so we need to raise the table by 0.75mm and remove this from each side. We do need a bit of clearance so I would remove about 0.9mm leaving a little bit of clearance at each side.

We also need to finish the thickness of the T-head, so measure this as well and move over the required amount taking the thickness of the tee about 0.2mm thinner than required. You now have a bar of the

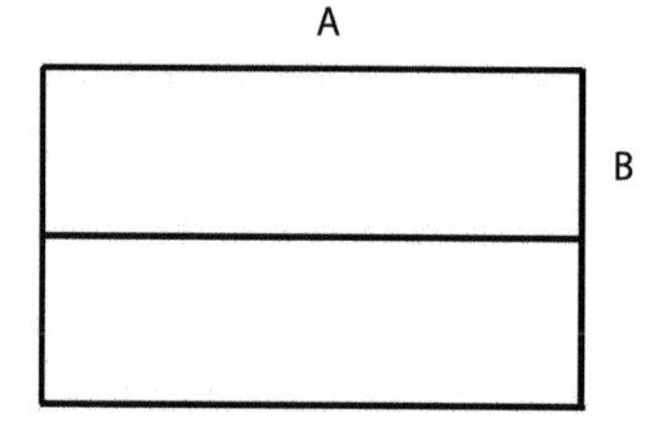

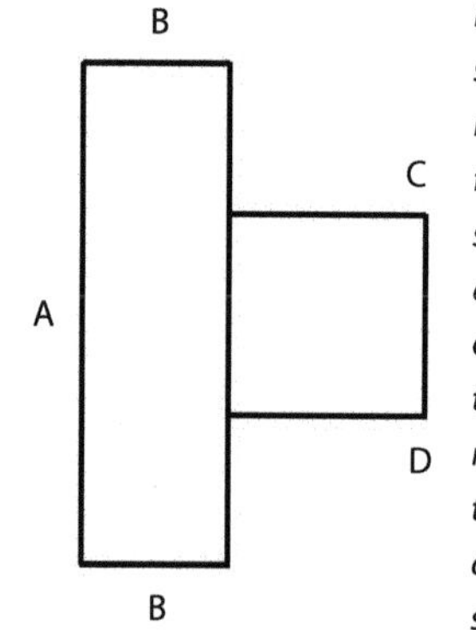

Fig. 7.5 This sequence shows you how to machine T-nuts. Machine faces A and B at the same setting then rough out C and D. Measure the width over C and D. Raise the table half the distance required to mill the tongue to size. Finish C, turn over and finish D at the same setting.

correct shape of the tee slot with a little bit of clearance. Machine the top of the bar if required so it is below the level of the top of the table.

Pitch out the holes along the bar, drill and tap them. Finally hacksaw the bar into individual tee nuts. Do not tap the T-nut right through, just leave a little bit of the thread undersize. Then the studs will not screw right through the T-nut but lock at the bottom of the thread. This will stop any tendency for the stud to force the T-nut up in the T-slot, possibly damaging the table.

ACCURACY IN MACHINING

Basic technique

It is not difficult to machine work accurately to thickness on a milling machine; it just needs a repeatable location to put the component in, and care used while machining. Even for the relative beginner it is easy to work to the nearest thousandth of an inch, as long as you have accurate measuring equipment.

The secret is to tap the work down onto the parallel so it is flat. Do this either with a copper mallet or put a bit of brass, copper or aluminium on top of the component and tap it down with a hammer. Take a cut from the top of the work, take the work out of the vice and measure it.

Then tap the component back down in the vice and take off the required depth of cut (more than one cut is allowed) until the correct size is reached. You will quickly find it is very easy to work accurately and to size using this method.

Machining accurate widths

Often you will have to machine a component such as a piece of sheet metal to an accurate width.

Clamp the metal on to the milling machine table (use a piece of scrap material or a piece of wood under the metal to protect the table) and put a cutter of known size in the spindle. Machine one edge to size, winding it towards the metal in the Y direction until it touches the edge of the work, wind backwards and forwards in the X direction making sure it is cleaned up.

Set the machine's dial to zero. Assuming we want to mill the piece of sheet metal to 100mm wide, move the machine table over by 116mm in the Y direction (100mm metal width + 16mm cutter diameter), lock the machine table, and clean up the second edge. To eliminate backlash you must turn the handwheel in the same direction as when you machined the original side; if you go past the 116mm, wind the table back a couple of turns and then wind to the 116mm reading on the handwheel. (We will look at backlash again later when we discuss pitching out holes.)

Using a depth micrometer

A depth micrometer is very useful when cutting slots or steps to size. The anvil on a depth micrometer is usually only 3 or 4mm in diameter so it will fit into most slots and steps you are likely to cut.

WHAT IS BACKLASH?

Backlash is the wear in the machine's feed screws. When you wind the handwheel to zero in one direction, the table is in one set position. When you wind the table to the same zero from the other direction, it will not be in the same position as when you wound it in the first direction. The difference in these two positions is the backlash. This will only be a few thou at most, but it is still a nuisance. So, when accuracy is important, always wind the handwheels in the same direction even if you have to wind it back a few turns and then wind it back to zero again.

Machining shims

Occasionally you may be called on to make some very thin shims. They may only be 10 thou or so thick. The easy way I have found to make these, assuming the shim has at least one hole in it, is to carefully cut the shims oversize with a pair of tin snips. Then clamp a piece of aluminium sticking out from the vice. Put the shims on top of the aluminium, put another piece of thicker aluminium on top and clamp the lot together with a couple of toolmaker's clamps.

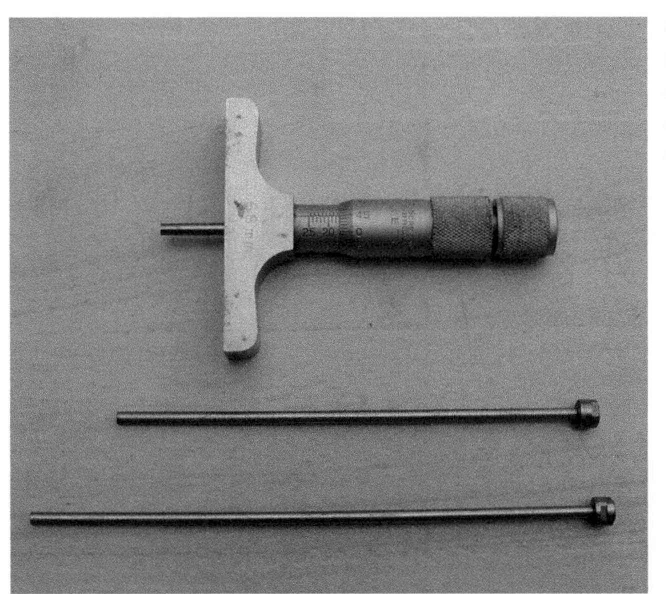

Fig. 7.6. A depth micrometer comes with interchangeable rods so it can be used for various depths.

Fig. 7.7. Machining thin shim is not difficult if you sandwich it between some scrap metal.

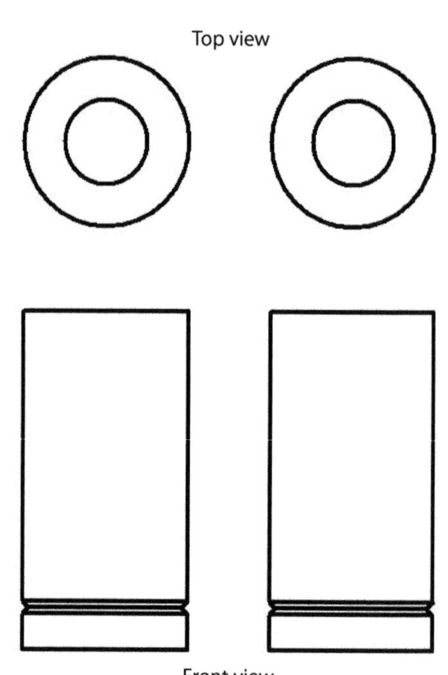

Fig. 7.9 Cylindrical squares will hold work at 90 degrees to the table's surface. The small groove at the lower end is optional but is useful so you know which end is true to the diameter.

Fig. 7.8. Although these circular shims were turned on the lathe, the principle is the same as milling square shims on the mill.

Carefully drill the required holes in the shims and fit nuts and bolts through the entire package and tighten them up. Remove the toolmaker's clamps and you can now machine round the outside profile of the shims. Undo the bolts and remove the finished shims.

Parallels bolted down

You can make a nest on the machine table by bolting parallels down to the table. Some parallels have holes through them so you can put bolts through them while others will have to have clamps on them. Clock the parallels with a dial test indicator if they need to be parallel to the table, or use a protractor from the edge of the machine table if they have to be set on an angle. The component can then be pushed into the corner made by the parallels and secured with a clamp.

Cylindrical parallels or squares

Cylindrical parallels or squares are, as their name suggests, cylindrical in shape with a flat turned face at exact right-angles to the upright side of the diameter. You can bolt these parallels directly onto the machine's table by using a bolt right through them. They are useful for all sorts of machining, from squaring up the ends of angle to machining angle plates, and you can also clamp blocks to them if you need to square up something.

They are easy to make. Just machine the outside diameter and the end face at the same time. Put a small groove at the lower end next to the machined end, and you will always know which is the square end that goes closest to the table.

You can also bolt them onto an angle plate, either parallel to the machine's table or at different heights to set the workpiece on an angle.

Cylindrical square as a vice stop

You can bolt a cylindrical square at the end of a vice to use as a vice stop. It can be fixed as far away from the vice as the machine

Fig. 7.10. This pair of 1 2 3 blocks will find lots of uses on the milling machine.

will allow for machining the ends of long workpieces.

1 2 3 blocks

1 2 3 blocks are so called because they usually have a ratio of 1, 2 and 3. The longest length could be 3in, the width 2in and the thickness 1in: hence 1 2 3. Larger 1 2 3 blocks are available called 2 4 6 blocks, which are twice the size of 1 2 3 blocks. The blocks are precision ground on all six sides and usually have a selection of through holes and tapped holes so they can be bolted together. They are very useful when used for workholding.

ANGULAR MACHINING

Angle plates

Two angle plates bolted together can be very useful in locating components, especially strip that needs to be drilled in the ends. You bolt the two angle plates together so their faces are offset, which leaves a step to drop the component into. You can clamp two angle plates to the table with a component clamped between – perhaps a casting, to have its face machined. You could also bolt a block of metal on an angle to an angle plate to create an angle machining fixture.

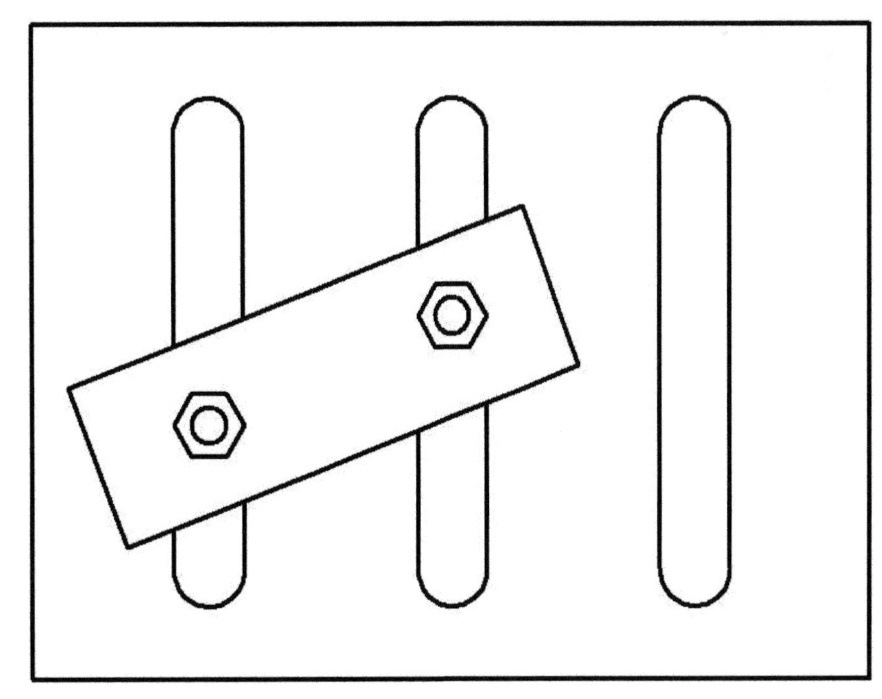

Fig. 7.11 A metal block mounted on an angle plate can be used to set work on an angle.

Machining angular faces

Slipper block

A method I have used to set up or mill an angle on a component is one I have not seen described before. I have used it successfully to set up a machining fixture to within ±2 seconds. (Angular dimensions run degrees, minutes and seconds.) I set up the fixture and it was spot-on first time. What you have to do is work out the drop for an angle over a specific length and then machine a block to this length and depth. Then, after very lightly deburring the block, turn it over in the vice and machine off the bottom of the block, which will be at the required angle. The resulting block is usually called a slipper block. These blocks are ideal especially for shallow angles.

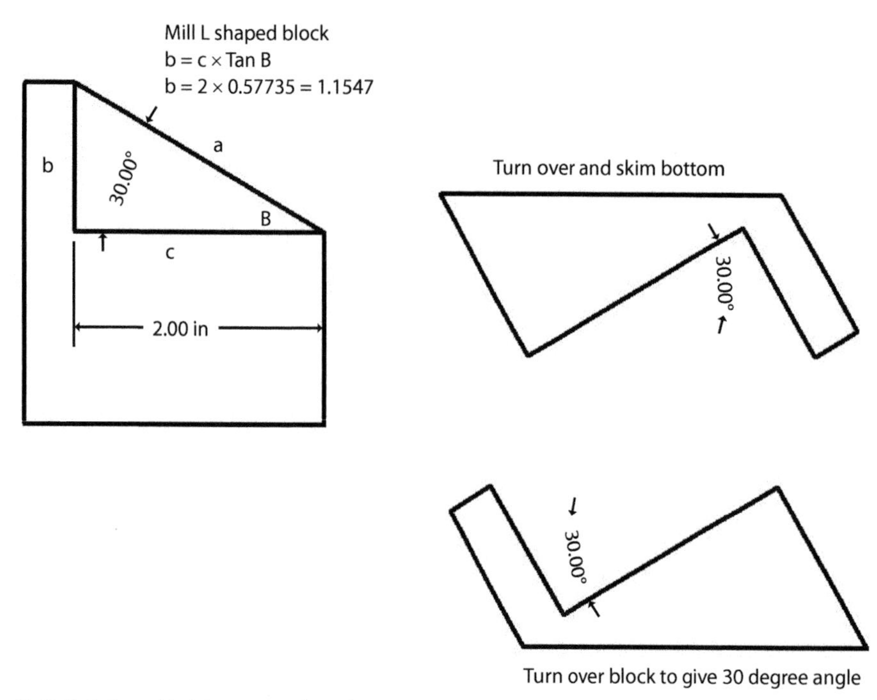

Fig. 7.12 A slipper block is easy to make and is very accurate.

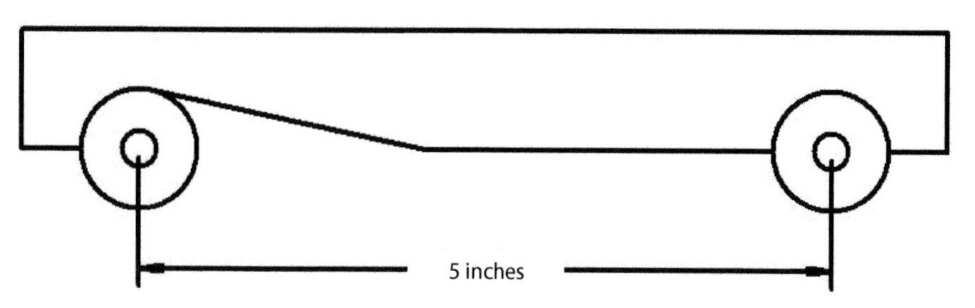

Fig. 7.13 A sine bar consists of a length of bar with two exact size disks mounted on it at known centres.

Sine bar

You can also use a sine bar in the vice to set work on an angle. You will need to refer to trigonometry tables to set up the sine bar. This is not difficult to do. Assuming your sine bar is 5in long (the 5 inches will be from the centre of one circular end to the centre of the other circular end) you just need to multiply the sine of the angle required by the length of the sine bar and that gives you the size of packing you need under the sine bar. You do not need a set of expensive slip gauges to use as packing – just make up the required thickness with pieces of bar and shim, or turn a piece of round bar to the required diameter.

Tilting vices

Tilting vices, mentioned earlier, are usually used for light drilling work on the drilling machine but there is no reason why you could not use one for light milling on the mill.

Adjustable angle plates

This type of angle plate can twist from 0 degrees, when the moving face is horizontal

Fig. 7.14. This angle plate is adjustable. Although small, a machine vice can be mounted on it.

Fig. 7.15. The scale on the end allows you to set the table on the correct angle. For exact angles, you could always set the angle with a sine bar.

to the machine's table, right round to 90 degrees. The angle plate can be set to any angle between 0 and 90 degrees and locked in place. This type of angle plate is usually big enough to have a small machine vice mounted on top of it.

Compound angles

Compound angles are where the workpiece is angled in two planes rather than just one. You could machine a compound angle by putting a small vice at an angle on the tilting angle plate or by putting the vice at an angle on the milling machine's table and tilting the machine's head on another angle.

Angular set squares

A cheap source of angle gauges is the plastic 30, 45 and 60 degree set squares found in school geometry sets. They are very accurate and if damaged can be replaced cheaply.

WORK STOPS

Vice stops

There are many ways to fit a work stop to a machine vice. We have already discussed fitting two vices to the milling machine and using a cylindrical square as an end stop. Another simple method is to put a magnet onto the vice jaw. This makes quite a good stop although steel swarf has a tendency to stick to the magnet. We could also drill and tap the vice's side and screw a block onto it.

Holding a round bar

A useful method I have used in the past for round bars is to mill a groove all the way along the fixed jaw of the vice to take the round bar. Then drill and tap a small hole in the top of the fixed vice jaw and use a clamp to clamp a bit of round bar into the end of the groove. Then you can put the round workpiece in the groove and slide it up to the stop and nip up the vice.

Another variation of this method is to put a dowel pin in the fixed jaw and use this as a fixed stop.

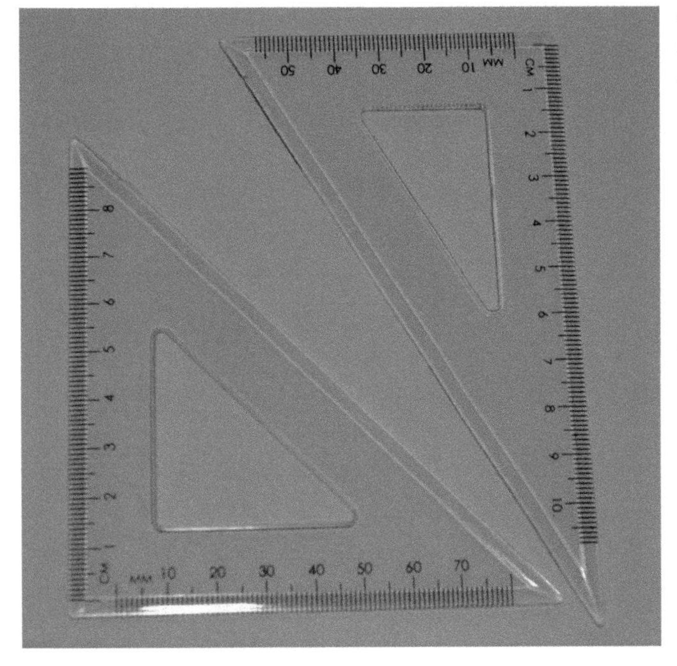

Fig. 7.16. Plastic set squares are very cheap and very accurate.

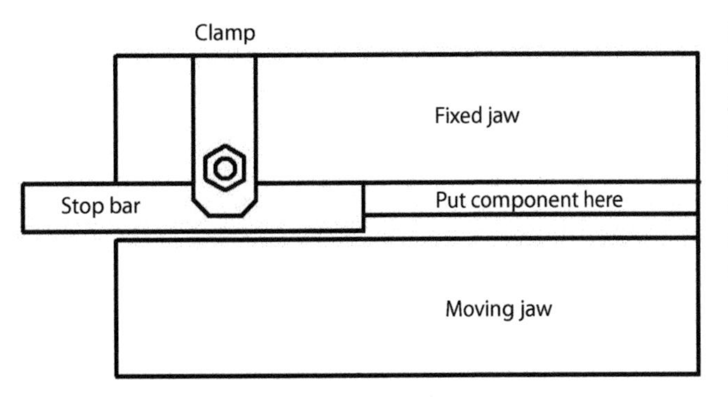

Fig. 7.17 Clamp a stop to the machine vice for production work. Push the work up to the stop and tighten the vice. The component is held in the same position for the whole batch.

PROFILING A COMPONENT IN THE VERTICAL PLANE

One of the most useful operations you can perform on a CNC milling machine is profiling a component in the vertical Z plane. There are two options to this, profiling in the ZX planes and profiling in the ZY planes. This technique requires a computer with a CAD package to work out the co-ordinates. It may take a little while and a bit of thinking to work out how to do this, but persevere and you will be able to do something that many professional millers do not know how to do. This profiling technique is normally used with CNC machines, but there is no reason why we cannot use it on a manual machine; it just takes a bit longer.

Ball nose cutter method

It is possible to do profile milling one step at a time with a ball nose cutter. Study the method and all should become clear. This is probably the most difficult thing to get to grips with in this book but it will be worthwhile if you do.

- *Using the CAD package draw the profile you wish to produce.*
- *Then you need to draw the outline outside of the existing drawing, but it must be an exact distance offset from the original profile. This distance will depend on the size of the cutter that will be used to cut the profile. The cutter used will be a ball nose cutter (sometimes called a bull nose cutter).*
- *For the purposes of this exercise we will use an 8mm ball nose cutter and the offset distance will be exactly half the ball – that is, 4mm offset. So we draw the profile and another profile exactly 4mm larger than this first profile.*
- *We will machine an 8mm radius on the top of a block, then the angle and finally a 4mm radius. (You could machine any shape or even a 45 degree angle across the corner of the workpiece.) The limiting factor is the 4mm radius we have chosen; we could have used a smaller cutter but then we would lose strength in the cutter but for an external profile the cutter diameter is immaterial.*
- *Once we have drawn our profile and the tool path of the ball nose cutter, we have to calculate the co-ordinates. We will take a step depth of 0.5mm to machine the profile.*
- *Then work out the co-ordinates on the CAD package and write them down on a piece of paper. Write all the co-ordinates down and tick them off as you do them so you do not make an error in the machining.*

The centre of the cutter is the centre line of the 20mm radius and 4mm above the top of the workpiece. This is the 0 0 datum position.

The position of the next cut is 0.5mm deep and the vertical offset distance is the position of the centre of the ball at 0.5mm deep.

The next cut is offset by the position of the ball centre at 1mm deep. All you have to do is keep moving the cutter over to the centre line of the ball and to the depth of cut.

You can do any profile you like using this method. You can even machine a profile around a workpiece in the XY plane (flat on the table or in a vice) by using a known diameter of cutter and just plunging the material away. You could even cut a hole using the same methods. If you need a smaller radius in the corners, just drill small holes where you need them before drilling and cutting the profile.

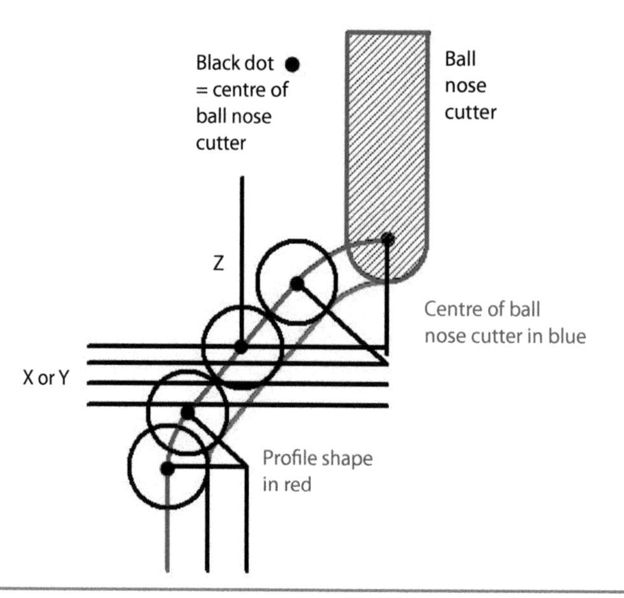

Fig. 7.18 It is possible to do profile milling one step at a time with a ball nose cutter.

8 Machining on the Table

Some jobs are too big to hold in the machine vice so you will have to clamp them down to the milling machine's table.

PROTECTING THE TABLE

When the job is clamped down, you should have some sort of packing underneath to protect the table from cutters or drills marking the table. Either clamp a parallel or a strip of metal under the work or clamp the work on a bit of MDF or plywood. A piece of MDF or plywood on top as well will also cut down on vibration. If the job is really big and difficult to clamp down perhaps it has some holes or a cutaway that you could put some studs through.

If the job is quite long, rather than clock it up push it against the dovetail slide, if the milling machine has one, to line up the edge of the work with the edge of the table. Once the work is clamped down, you may be able to mill both sides of the work parallel at one go. This is especially easy on a large turret milling machine where you can move the ram in and out to get at both sides of the work.

If the workpiece is too wide, you may be able to machine one edge and most of the two ends at the same time; then flip the workpiece over to finish the other side and the remainder of the ends.

With a bit of forethought, quite large workpieces can be held on the machine's table with no more than a bit of packing underneath and a couple of clamps.

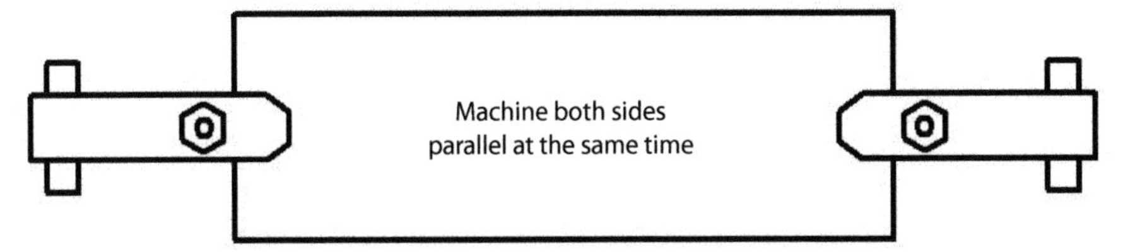

Fig. 8.1 Both sides of this metal sheet can be machined at the same setting. This ensures the sides are parallel to each other.

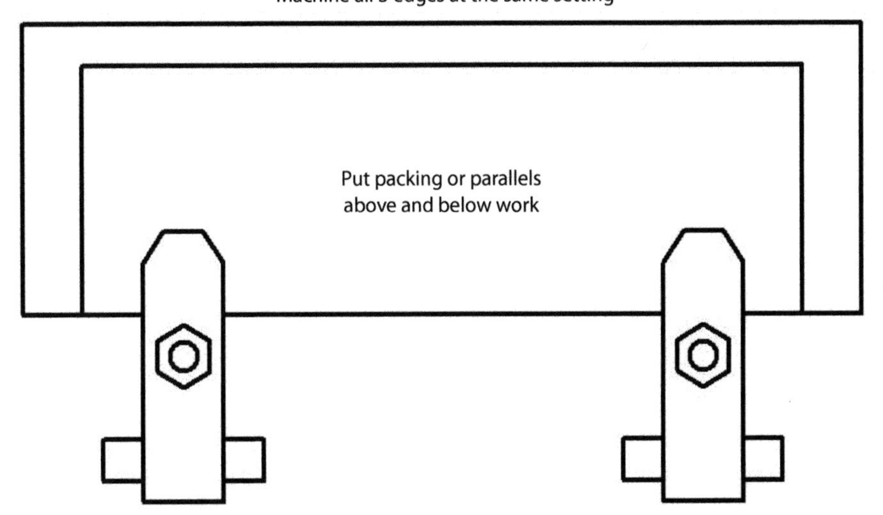

Fig. 8.2 Three sides of this component can be milled at one go.

WORKPIECE HOLDERS

Tenons

Tenons are small blocks that fit into the top of a table's T-slots. They should be a good fit, not loose but not so tight that you have to bang them in. You may have tenons on the bottom of your vice, dividing head or rotary table.

You can also have loose tenons that fit into the T-slots that you put the work against to line it up with the table.

If you want, say, a row of holes being drilled to line up with a T-slot or the edge of the work being machined to hang over a tee slot, you can put a parallel against the edge of the tenons to space the work into the correct position.

If you want to put an angle plate onto the milling machine parallel to the X axis, put a couple of tenons into the T-slot and push the angle plate up against the tenons to save clocking up the angle plate. Tenons do not have to be square; they can be rectangular or even be made from round bar.

T-nuts

T-nuts are used to clamp things down to the machine table. They are so called because they resemble an inverted T. They should be a good fit in the table's T-slot. T-slot nuts can be purchased from the usual suppliers but they are also very easy to make.

Under no circumstances should you use hexagon headed bolts or nuts as T-nuts. They are not suitable and could pull out from the T-slot, possibly damaging the table at the same time. T-nuts should have parallel sides to offer maximum strength and clamping support.

One place it is permissible to use a rounded profile to the top of a T-nut is on angle plates where the angle plate slot is rounded at the ends. The rounded profile enables the T-nut to slide right up to the edge of the angle plate's slot.

Fig. 8.3. Top hat nuts have a collar to spread the load.

Studs

Studs screw into the tee nuts to clamp down work, angle plates, vices and so on. You will need a selection of studs of various lengths. A simple way to obtain studs is to buy some threaded rod called studding or all-thread. This is often supplied in metre lengths which you can cut down as needed.

A couple of lengths will keep you in studding for quite a while. Cut to length and chamfer the ends. A quick clean-up with a triangular file will clear the burr from the end of the thread.

Nuts and washers

You will need nuts and washers to go onto the studding to clamp things down. Ordinary thin washers will not be much use for this purpose. Get some round bar and use the lathe to make your own washers. The diameter should be chosen to suit the size of your vices and angle plates. You want to completely cover the clamping holes and slots with at least 6mm overlap, preferably 10mm or more. Make them at least 5mm thick up to about 8mm, so they are solid enough to take the clamping pressure.

Ordinary nuts will be fine but if you can, get shouldered nuts. These are like ordinary nuts but have a collar below the nut to spread the load. These nuts can be purchased from specialist machine tool dealers.

Clamps

You will need a selection of clamps to hold work down to the table and packing blocks to go under the end of the clamps. You can buy sets of clamps that have angular teeth on their ends to match angular packing blocks. By raising the back of the clamp and putting a packing block under, the height of the clamp can be varied.

Fig. 8.4. Clamps should be ⅓ at the clamping end and ⅔ at the packing end, which should be slightly higher than the front of the clamp. Notice the teeth on the packing block and clamp.

It is important that the packing block is used correctly. The stud should be about ⅓ of the way along the clamp from the work. The remaining ⅔ of the clamp should be behind the stud at the packing block end. The packing block end should be slightly higher than the work end of the clamp so the top of the clamp runs very slightly downhill to the work.

A selection of flat clamps and packing blocks will be found useful as well.

A different type of clamp is the swan-neck clamp, so called because it resembles a swan's neck. This sort of clamp is ideal when you want to keep the clamps low down to the table, perhaps to avoid fouling on the cutter. A refinement of the swan-neck clamp is one where the clamp is fitted with a pivot pin; the clamp's back end clamps directly onto the table, and the clamp itself clamps on to the work.

Fig. 8.5. This clamp rests on the work or fixture at one end and on the table or a bit of packing at the other end.

Sometimes you will want to stop the clamp from dropping onto the table when you release it; perhaps you have to spin the clamp at 90 degrees to free the work. Either put a nut and a washer on the studding below the clamp or use a spring and a washer to support the clamp. Make sure the washer is free when you tighten up the clamp.

CLAMPING ROUND SHAPES

Hollow components

Hollow components must be supported when clamping them otherwise you may crush them. You could put a bit of round or square bar into a tube to stop it from squashing.

Using vee packing blocks

It is useful to have a few vee packing blocks to support round bars and tubes while clamped down on the table. Vee packing blocks are easy to make; just put a block of metal into a vice at 45 degrees and mill the groove with a standard cutter.

Round bars

A large round bar can be clamped to the milling machine's table by dropping it into the tee slot. The bar should be at least 50 per cent larger than the width of the tee slot. Do not over-tighten the clamps; just use sufficient force to stop the bar from moving. You could add a clamp on the table at each end of the bar to stop the bar sliding along the tee slot when you are machining it.

CLAMPING UNUSUAL SHAPES AND SIZES

Using toolmaker's clamps

It is possible to clamp a toolmaker's clamp on the machine's table or in the vice and use it to clamp workpieces down, especially small or funny shaped workpieces. A selection of different sizes of toolmaker's clamps can be made in the home-workshop or purchased from your tool supplier and will be found very useful.

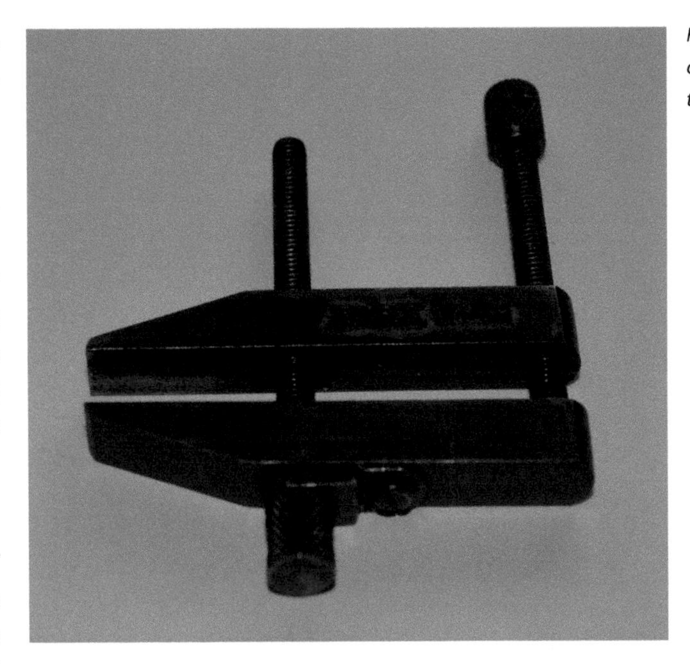

Fig. 8.6. Toolmaker's clamps have many uses in the workshop.

Clamping castings

You should use a piece of soft packing (such as thin cardboard) between the rough casting and the surface of the angle plate or machine's table to stop indentations from the casting ruining the machine or fixture.

Clamping a casting often requires various thicknesses of packing and shim while being clamped down. Try not to distort the casting as it may not be flat when the clamps are removed.

Fig. 8.7. This casting has brass packing and a steel rule to support the casting for machining the bottom.

Fig. 8.8. After machining the bottom the casting is turned over and clamped to the milling machine table to machine the top parallel.

USING AN ANGLE PLATE

Milling parallel on an angle plate

If you bolt two discs onto an angle plate and clock them so they are parallel you can clamp the work true onto the angle plates while resting the work on the disk. Then you can machine the work parallel and to size. This method is useful when you are just starting out and may not have a machine vice. It will also be useful for long workpieces that cannot fit into a single vice.

Box angle plate

Another type of angle plate is the box angle plate. This is like an open ended box with slots all round. You can use it like an ordinary angle plate or clamp work on top of it to raise the work off the table. It will find many other uses in the workshop.

Fig. 8.9. This is a small box angle plate.

Milling dovetails

It is possible to mill a long dovetail by clamping the work down to the machine's table. It may be possible to tilt the machine's head to mill the dovetail but if this is not possible, you can buy dovetail cutters and inverted dovetail cutters to do this. Dovetails are normally 45 or 60 degrees. If cutting a dovetail on both sides of a workpiece try and machine both sides at the same time to ensure the sides are parallel to each other. Shorter dovetails can be machined in the machine vice.

USING CHUCKS

The three jaw chuck

The three-jaw chuck will be found to be very useful on the milling machine table. Once clamped down you can hold all sorts of round and hexagonal components for milling and drilling. Ideally you will turn the workpiece on the lathe and then transfer the chuck with the work still in it to the milling machine for working on. This ensures accuracy and concentricity.

However, this method is not usually possible when you are making a batch of similar components. In this case you will have to do all of the turning first and then set the chuck up on the milling machine to mill the components as a batch.

You can use either the chuck's internal or external jaws to hold the work or an alternative way is to use soft jaws. The soft jaws can be bored, milled or otherwise formed to take the workpiece. Packing can be used on the hard or soft jaws to raise the component clear of the hardened jaws for drilling or milling.

Fig. 8.10. Chuck on milling machine table. This has standard drill jaws.

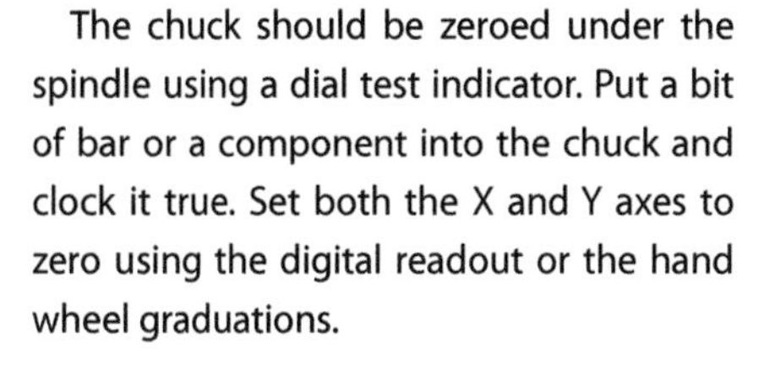

The chuck should be zeroed under the spindle using a dial test indicator. Put a bit of bar or a component into the chuck and clock it true. Set both the X and Y axes to zero using the digital readout or the hand wheel graduations.

Milling complete components

Sometimes a component is too small to hold in a vice or chuck to machine it. In this case, take a larger piece of material and hold it in the chuck or vice. Then you can machine the item complete all over and finally cut it off from the parent bar.

Fig. 8.11. Unmachined soft jaws. You can machine any shape you like into these jaws.

Fig. 8.13. This block has been machined on both sides, the top and both angles all at one setting. The slot will be machined next and the block will then be turned over for the bottom face to be machined.

Fig. 8.12. Chuck fitted with external jaws. The backplate is threaded for the lathe but is the same bore and thread as the dividing head.

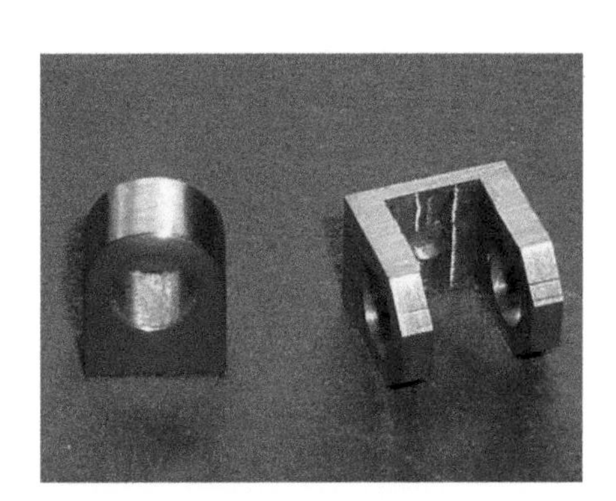

Fig. 8.14. The finished block is on the right waiting for cleaning up. It is about 10×10×10mm square.

Here are some PCD charts for you to pitch out holes up to twelve divisions using linear co-ordinates, although you can use a digital readout with a PCD facility if you have one.

Not all numbers are covered because you can use some of the six-hole co-ordinates to do three holes, and drilling four holes on centre is just a case of moving left and right and forward and backward by the PCR (pitch circle radius).

PITCHING OUT HOLE CIRCLES

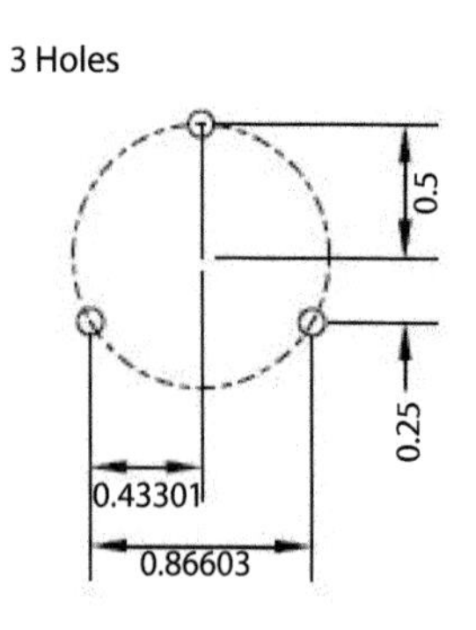

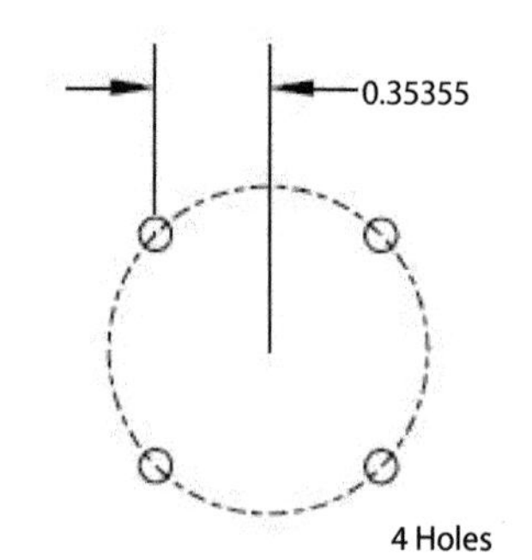

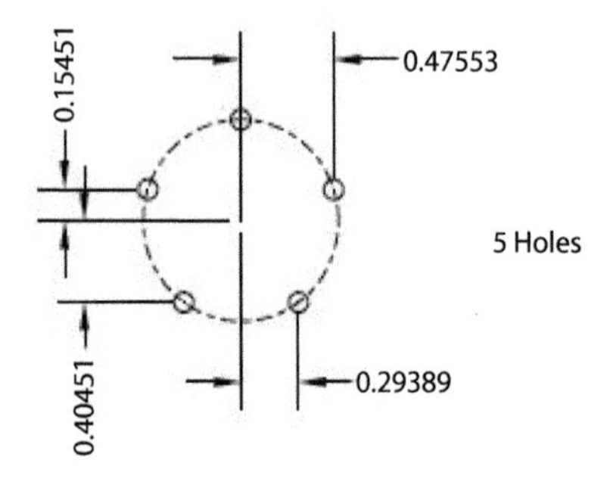

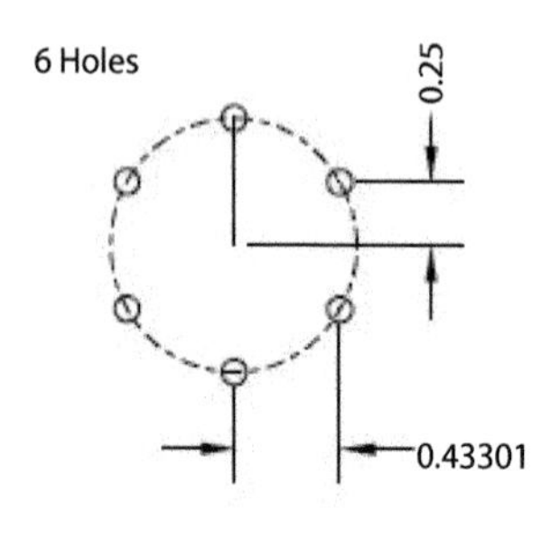

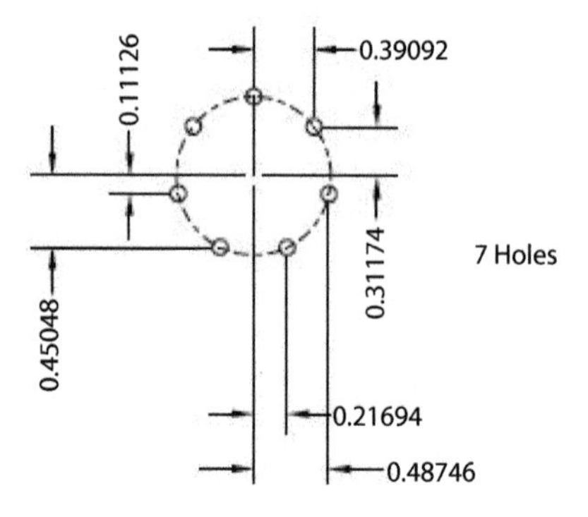

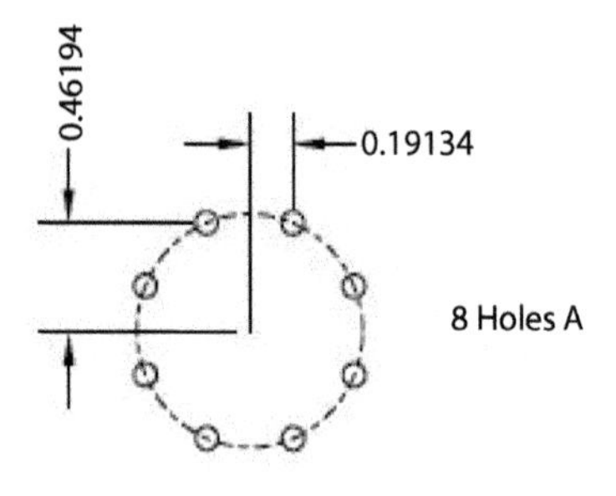

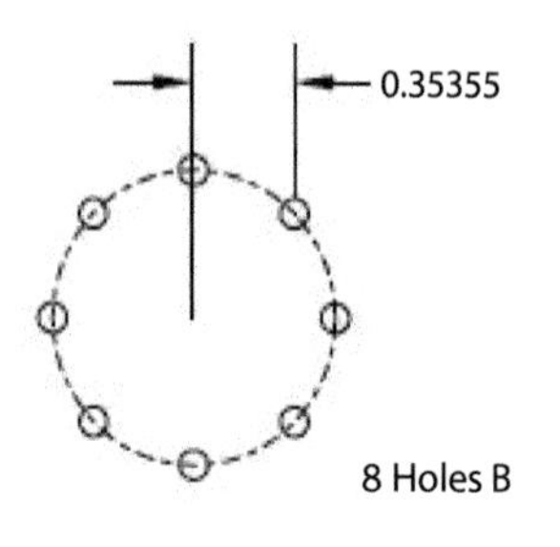

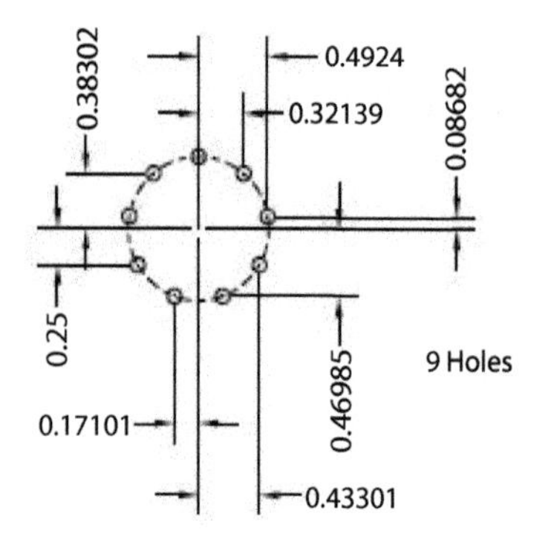

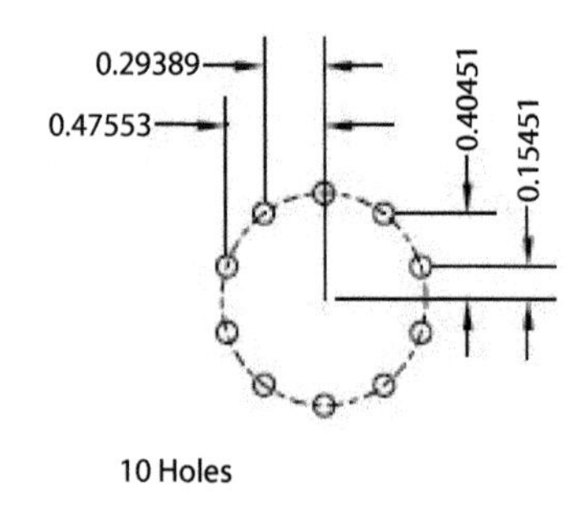

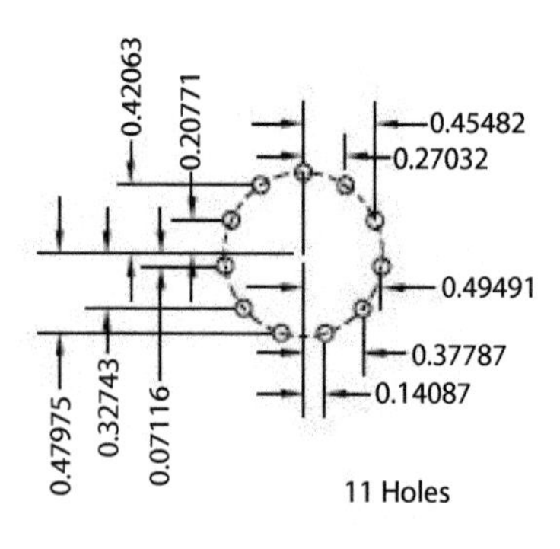

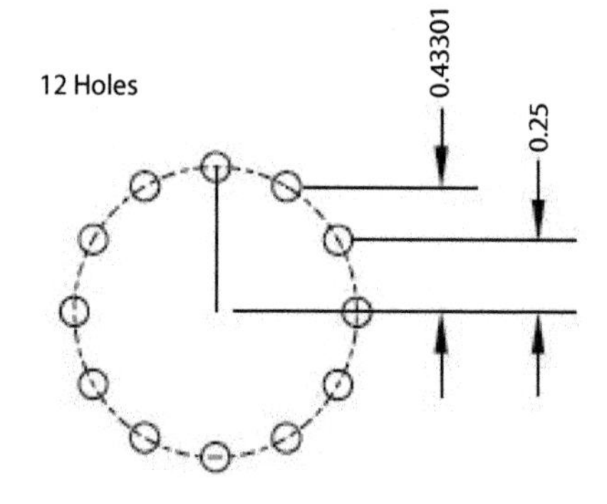

Fig. 8.15 Co-ordinate drilling charts for 3, 4, 5, 6, 7, 8, 9, 10, 11 and 12 hole centres.

It is also useful to pitch out 24 holes into a chuck backplate for dividing in the lathe.

Fig. 8.16. This chuck backplate has been drilled in 24 places to allow dividing on the lathe.

Fig. 8.17. The chuck, minus its jaws, is clamped down to the milling table with a bolt through the centre and clocked up under the spindle. Any chuck can be clocked the right way up using this method.

Fig. 8.18. The back plate was centre drilled using co-ordinates.

Fig. 8.19. Then the back plate was double drilled to the size required. Double drilling is where you drill an undersized hole and then follow through with a size drill.

Using chuck spiders

A chuck spider is machined to clear the chuck jaws so the component can rest on it in the chuck. It is basically a disk with three slots machined in it to take each of the chuck jaws. You will find spiders are of use a lot especially when machining thin items.

Another method of supporting work in the chuck is to use three bits of round bar under the work. If you turn grooves at each end of the bars you can clip a single tension spring on both ends and use the spring to hold the bars on the chuck jaws.

The collet attachment

The collet chuck is much better for small work than a three-jaw chuck as it grips all around the work to support it. You set up the collet attachment on the milling table in a similar way to the three-jaw chuck. You may have to fit a backstop inside the collet chuck to stop the work pushing back in the collet. The backstop can be as simple as a bit of tee shaped bar placed inside the collet and resting on the machine's table or the bottom of the tee slot.

You can either clock the collet chuck or clock the work diameter. In this case I clocked the actual work diameter here on each individual component as the spigot I was holding on was not true to the outside diameter. It only took a few seconds to do each component with the guarantee that the holes were true to the outside diameter. After the holes were accurately pitched out, the component was held true against a circular pocket in the lathe, using a running centre and a pressure pad, and the outside diameter and face of the components were cleaned up.

Fig. 8.21. The holes were pitched out in the X and Y directions to give four holes.

Fig. 8.20. This ER collet attachment can be clamped down to the milling machine table. You can either clock the collet chuck or clock the work diameter. I clocked the actual work diameter here on each individual component, as the spigot I was holding on was not true to the outside diameter. It only took a few seconds to do each component with the guarantee that the holes were true to the outside diameter.

The four-jaw chuck

Irregular castings can often be held in a four-jaw chuck on the machine's table for machining. Just clamp down the four-jaw chuck and set up the casting.

Chucks and the angle plate

Any of the previously described chucks can also be clamped at right angles to an angle plate for further work on the sides of components. A typical example of this is to use a slitting saw to cut a slot in the head of a threaded component, to take a screwdriver or to drill a cross hole in a bit of bar.

The Keats vee angle plate

Although the Keats vee angle plate is normally used on a lathe's faceplate, there is no reason why you cannot use it on the milling machine's table or mounted on an angle plate. Do consider this method, as you will find it useful.

Fig. 8.22. A Keats vee angle plate can be used on the milling machine as well as shown here on the centre lathe. It can be mounted flat on the table or at 90 degrees using an angle plate.

9 Using a Rotary Table and Milling Radii

You will find that a rotary table is a very useful piece of workshop equipment. It can be used to mill circular components that would otherwise be impossible to machine, to drill equidistant holes on a flange, to mill larger holes that would not be possible to drill, to cut straight lines at angles to each other and to cut arcs. The rotary table usually has a worm and wheel drive although there are some small tables that are rotated directly by the use of a bar inserted into a hole in the outside diameter of the table.

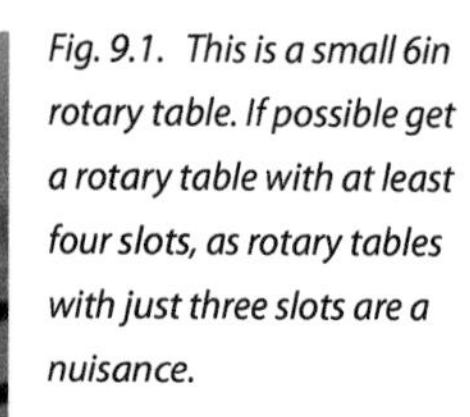

Fig. 9.1. This is a small 6in rotary table. If possible get a rotary table with at least four slots, as rotary tables with just three slots are a nuisance.

Fig. 9.2. Rotary tables can usually be used vertically as well as horizontally.

DESCRIPTION OF THE ROTARY TABLE

Clamping the work

The top of the table usually has three or four (sometimes more) T-slots that are used to clamp the work down. The chances are, the T-slots are smaller than you use for the milling machine table so you will have to make a few T-nuts to fit. You might as well make a few small clamps to clamp the work down with as well.

Centre hole

The centre of the table may have a precision finished hole in the centre of the table for clocking up the table's centre or for locating work. The hole may be a parallel hole, although most of the small rotary tables you are likely to come across will have a shortened No.2 or No.3 Morse taper hole.

If you have the Morse taper version, I suggest you plug a blank Morse taper arbor into your lathe headstock and drill and bore a known size parallel hole in it, not through, and use this in the centre hole although you may have to shorten the back end of the Morse taper. Then you can make various plugs to suit the known hole size on one end and a diameter to suit the work on the other end.

Protecting the table

It is important to keep the top surface of the rotary table clean and free from dents and machining marks. For this reason, the work should not be clamped directly onto the rotary table but should be raised above the table's surface by some sort of packing piece. Something soft such as aluminium or brass would do to protect the rotary table's surface

CENTRING

Centring the table

To centre the table below the machine's spindle, you need to put a lever dial test indicator into a collet or drill chuck that is in the machine's spindle. Move the rotary table approximately under the centre of the spindle. You can do this either by sliding the rotary table into position and clamping it down or by clamping the table down first and then winding the milling machine table's handles so the rotary table is under the spindle.

Bring the dial test indicator down into the rotary table's centre hole and rotate the spindle carefully. By using the milling machine table's handles you can now centre the table exactly under the spindle. You must do this by carefully rotating the dial test indicator in the machine's spindle. You cannot do this by rotating the rotary table's handwheel. If you just rotate the rotary table, the centre hole will remain in contact with the dial test indicator and you cannot set it true this way. Many beginners do not realize this and then wonder why the rotary table is not under the spindle's centre line.

Centring the work

Once the rotary table is centred and the digital readout or the machine's handwheels have been set to zero, or the position required, we can set the work true on the table. A simple way to do this is to put a round bar into the machine's spindle and line this up with a hole in the workpiece. If you do not have a bit of bar the correct size to fit into the hole, turn a bit of bar on a taper that will go into the hole about halfway and use this to line up the work. After a while you will have several different sizes of taper you can choose from.

Alternatively, turn a small pin to fit the rotary table at one end and the work at the other end. If the pin needs to be removed – say to mill the hole in the workpiece larger, and the hole in the workpiece is smaller than the hole in the table – put a sleeve into the rotary table's hole to reduce the diameter smaller than the hole in the work so the pin can be removed.

Say you want to mill a radius on the end of a bit of flat bar, but there is no hole in it. Put a pin in the machine's spindle with a drilled disk the size of the outside of the bar. Put the flat bar onto the rotary table and line up the workpiece with the drilled disk and clamp it down. The bar can now have the radius milled on it in the correct place, and you know that it will clean up.

If you have more than one workpiece to machine, you could set a side stop on the rotary table's surface so all of the workpieces can be milled at the same setting.

You can use this method to set up internal radii as well. Use the same pin and disk method, but set the bar so the end that needs the internal radius machined on it is just covered by the disk. Now you can machine the radius knowing for sure that it will clean up.

There are many more ways that you can set work onto the rotary table; they are only limited by your own imagination.

RADIAL WORK

Machining radial slots

Often you will be called on to machine a radial slot. This will usually be measured in degrees from the centre point at one end of the slot to the centre point of the other end of the slot. I would drill the centres of each end of the slot undersize first, making sure the angular movement was correct at each end. Then I would finish the holes to the correct size using a slot drill, end mill or boring head so the slot ends are correct to drawing. After that, I would remove most of the waste material from the slot by drilling and then milling undersize. I would then finally mill both sides of the slot to size using an undersize end mill, one side of the slot at a time as discussed for straight slots.

Drilling radial holes

You can also use the rotary table as you would a dividing head. By moving the dividing head round by the correct number of degrees, you can pitch out holes on a PCD or any other angular co-ordinates. You can get a complete set of dividing gear to fit some rotary tables (*see* Chapter 10 on using a dividing head on the milling machine).

Milling radiating slots

You can mill radiating slots (slots that run straight from the outside of the plate to the centre or part thereof) into a plate in the same way as pitching out holes but by winding the milling machine's table along the slots, indexing round the required number of degrees and milling the next slot and so on.

Machining a large radius

Perhaps you need to turn the outside of a large diameter such as a traction engine rim, but it is too big for the lathe. If you can mount the rim on a rotary table you can mill the outside of the rim rather than turning it. As long as you can set the wheel up central, it will be true with the outside of the rim.

OTHER METHODS OF USING THE ROTARY TABLE

Usually a rotary table can be mounted horizontally (flat to the table) or vertically (90 degrees to the table). It can also be mounted vertically in line with the X axis or the Y axis. You can also get tailstocks for rotary tables. A typical use for a rotary table in the vertical axis is to mill two faces parallel or at a precise angle to each other.

Thinning down components

Although this method can be used on any shaped component, I have found it most useful when finishing off components.

We have a pear-shaped component machined on a rectangular block of metal that has a flange that was left when we made the shaped component. The problem is how do we machine the back of the shaped component? You may think this is difficult, but all you have to do is hold the shape against the fixed jaw of the machine with the flange against the moving jaw. Then you can machine half of the work to the correct finished size using an end mill or similar. Invert the work in the vice and machine the other end of the component to the finished thickness. You now have a shaped component with the flange removed. Rub on a flat sheet of wet and dry paper to clean up the finish machined face.

Milling a radius in the vice

We have a short strip of metal with a hole in the end and we need to form a radius at the end of the strip. Put a bar through the hole, and drop it into the vice at about 1 degree below horizontal. Mill across the top of the bar where the radius is to go, lower the end to 2 degrees and mill across again. Keep doing this until the radius is formed right around the end of the bar. Deburr and clean up the radius with a file.

10 Using a Dividing Head

The dividing head is a very handy piece of workshop equipment, used to hold workpieces for drilling and milling radially. The dividing head consists of a main body, usually a casting, with a spindle through it to take a chuck or collets and a tailstock to support the work. The spindle is usually driven by a worm and wheel. There are various ratios of gears in use but the most common is a 40:1 ratio. The spindle may also have a direct indexing plate, most probably with 24 holes for direct indexing. If the dividing head has a 24 position direct driving plate, divisions of 24, 12, 8, 6, 4, 3 and 2 can be made without using the worm drive.

Fig. 10.1. This small dividing head has a 40:1 ratio, which means 40 turns of the handle will rotate the dividing head 360 degrees.

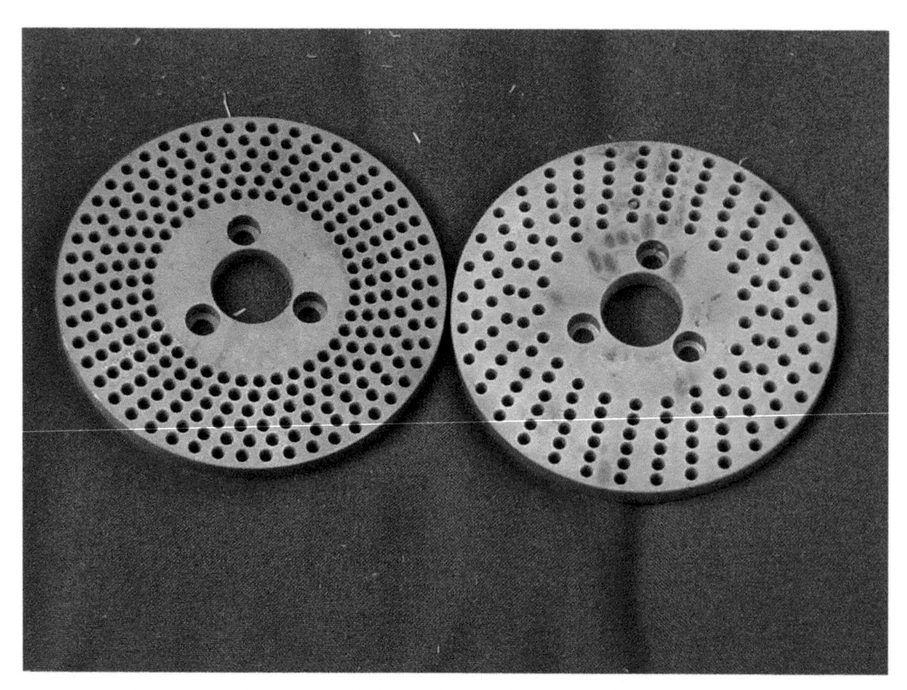

Fig. 10.2. Most dividing heads will come with three or four division plates as standard.

Fig. 10.3. Dividing heads are normally mounted to the left of the spindle and you turn the handle clockwise.

DISKS OR DIVISION PLATES

You can attach round disks (division plates) to the worm drive mechanism. These disks contain varying numbers of holes all set equally around the disk. Each disk can have several hole circles, each with different numbers of holes. The number of holes usually increases in number from the inner circle to the outer circle.

The disk is fixed to the main body, and a handle with a moveable plunger is fixed to the worm gear shaft so you can rotate the spindle through the worm and wheel. The plunger can slide in and out of the drilled holes on the division plate. Also mounted on the dividing head is a pair of sector plates. These plates can be rotated around the worm spindle. The sector plates can be set at various hole positions apart and are used as a guide as to how many holes the plunger can be rotated.

The number of holes

The number of holes the plunger should be moved is always plus one hole. This is because the plunger itself takes up a hole. For example, say you need to move the handle through 20 holes. As the plunger takes up one hole, if you move the plunger 20 holes, the plunger will only move 19 holes as it will be in the first and last hole; so you need to add an extra hole to the sector plates so the plunger will move through 20 holes and sit in the 21st hole. Not moving this extra hole is the biggest cause of scrap work when using the dividing head. I have seen experienced toolmakers make this simple mistake.

Sectors

To use the sectors, set the plunger in the starting hole. Set the holes plus one on the sector plates. Rotate the sector plate so the left sector is against the side of the plunger pin. Machine the work as required at this point. Rotate the plunger to the right until it is at the last hole on the sector plate (you may need to wind the dividing head an extra turn or two first) and lock the dividing head. Rotate the sector to the right until the left-hand sector is up against the pin. Continue until all the work is done at each position required.

DIVIDING A 13-HOLE DIAMETER

As a practical example, we will work out the way to divide a 13-hole diameter. Why 13 divisions? I have had to do 13 divisions in industry, and it is a real example that illustrates a lot of the problems encountered.

- ***We will assume a dividing head of 40 to 1, which is what is most likely to be found in the home workshop. One turn of the crank handle is 9 degrees.***
- ***360 degrees divided by 13 divisions = 27.69230769230769 degrees per division, so we have to turn the main crank three turns to get the 27 degrees which leaves 9 degrees that need dividing by 13 holes.***
- ***13 times 26 degrees = 351 degrees. That leaves 9 degrees = 1 turn, so any division plate that has a multiple of 13 holes will do to give us the odd fraction of the movement required. So if you have a 26-hole plate, move 3 turns and 2 holes to give the required movement. If you have a 39-hole circle, move 3 turns and 3 holes. If you have a 65-hole circle, move 3 turns and 5 holes and so on: 26, 39 and 65 are all multiples of 13.***

Do not forget to add the extra hole to allow for the plunger so 2 becomes 3, 3 becomes 4 and 5 becomes 6.

Ideally dividing should be done in one go, without any interruptions. Dividing is not difficult, it just requires care, absolute concentration and patience.

USING THE ACCESSORIES

Tenons

The dividing head is usually set parallel to the milling machine table. Often the dividing head will have a tenon under the body to locate in a table slot. Depending on the size of the table slot and the size of the tenon, you may have to make new tenons that fit the slot. You should be quite capable of doing this simple modification using the techniques you have learnt in this book.

Lining up the centre

To line up the centre of the dividing head, it is best to use a wobbler.

- Put a piece of bar into the dividing head's three-jaw chuck or preferably use a collet. Wobble from the bar and note the reading.
- Rotate the dividing head 180 degrees and note the reading.
- Move the table so the spindle is in the mean position of the two readings. You are now ½ of the wobbler and ½ of the bar diameter away from the centre line of the dividing head.
- Add these two dimensions together and that is how far you have to move the table so the spindle is above the centre line of the dividing head.

Aligning the tailstock

To set the dividing head tailstock parallel and at the correct height you need to put a dial test indicator into the collet or chuck. Fit the tailstock tenon into the table slot so it is close enough to the dividing head for the dial test indicator to touch on the tailstock's centre.

Use the dial test indicator to centre the tailstock at the correct height. If necessary, adjust the centre height of the tailstock making sure it is parallel as well as at centre height. Once the tailstock is at centre height and in line with the dividing head, you can move the tailstock backwards and forwards along the T-slot to accommodate any work that needs to be held between centres or between the tailstock centre and chuck or collet.

Tighten the chuck

On the cheaper dividing heads usually found in the amateur's workshop, the chuck will be threaded to fit on the spindle. This is not ideal as the chuck could possibly come loose. So screw the chuck onto the dividing head, place the chuck key into the chuck and give the key a slight tap with your hand to ensure the chuck is tight up on to the mandrel as far as it will go. There is no need to use a hammer to tighten up the chuck.

Now that the chuck is tight on the spindle, as long as we mill from the front of the table towards the back so the pressure on the chuck is in a clockwise direction, the chuck will tighten up rather than undo if it moves at all. If in any doubt that the chuck is sufficiently tight to resist movement, put a piece of scrap material in the chuck and take a milling cut from the front to the back to make sure the chuck will not rotate on the spindle.

SOME COMMON JOBS

Milling a hexagon head

One of the most common things you may be called on to do in the dividing head is to mill a hexagon head on the end of a bolt.

If the bolt has a parallel shank you may be able to hold the bolt in the chuck, but normally the bolt would be threaded and you will have to put the bolt into a threaded mandrel in the chuck. If this is the case, tighten the bolt up using a pair of pliers or mole grips on its head prior to milling the hexagon. It does not matter how tight you do the bolt up as long as it will not come undone. It will be easy to remove the bolt from the mandrel when you have milled the hexagon; just use a spanner on the newly formed bolt head to undo it.

To mill the hexagon, take a small cut on one side, rotate the dividing head 180 degrees and take another cut at the same setting. Measure the width across the two flats – this is the AF (across flats) dimension. Say we are milling a 10mm bolt head and the AF dimension we have just measured is 19.5 AF; the AF for a 10mm bolt should be 17 mm so we have to remove 2.5mm, or 1.25 from each side of the bolt head. Raise the table 1.25mm and mill all six flats, and we will have a 10mm hexagon bolt with an AF of 17mm. Job done!

We can use the same principles to mill any number of flats on any bit of bar held in the chuck, from a single flat to multiple flats.

Fig. 10.4. A gear cutter.

Milling dovetails

You can mill dovetails using the dividing head. Bolt a strip of flat metal onto a bar held between the chuck and the tailstock. This bar can be square, hexagon or round. If round, mill a flat on it sufficient to mount the flat bar on. The flat bar must be parallel to the machine in the table's X axis. The Y axis is not so important as long as there is sufficient material to clean up.

Rotate the dividing head so the edge of the flat bar is at the required dovetail angle and mill the dovetail. Rotate the dividing head so the flat bar is rotated to the other side of the dovetail and mill it. You now have a very accurate dovetail made with the minimum of effort and measurement. If you do not want holes in the main face of the dovetail slide, either make the dovetailed piece longer and cut to length after milling or use a clamp at each end of the slide to avoid the holes.

Milling a gear wheel

You use the same method to mill a gearwheel as you do to say mill flats on a bit of bar. The main difference is the shape and form of the cutter. The gear cutter will have multiple teeth and the gear blank will need indexing by the exact number of teeth in the gear. The gear cutter must be set at the exact centre height of the dividing head. To do this, use a height gauge to measure the difference from the top of the gear blank to the top of the gear cutter. This dimension should be ½ the bar diameter less ½ the cutter diameter assuming the cutter teeth are central in the cutter thickness. Set the height of the cutter using the height gauge.

The gear cutter should be on the far side of the dividing head and the gear cutter should be rotating in the same direction as a normal twist drill. The gear cutter should be moving through the gear blank towards the chuck. If you don't have a gear tooth vernier, make the gear blank diameter to nominal size, touch on to the outer diameter of the blank and take a depth of cut according to a tooth depth chart or a tooth depth calculation. Doing it this way should ensure the gears mesh correctly.

If you have many small gears to make you could put a long piece of bar, previously turned to the outside diameter of the gear, into the dividing head and supported by the tailstock. Then you can mill all the way along the bar to make a 'stick' of gear blanks. When you have machined the entire length, you can put the stick into the lathe, supported by the lathe's tailstock, and part off the individual gears. Finish the gears by holding in soft jaws and facing both sides of the gears. Finally drill and ream for the shaft. This is a quick method of making large quantities of gears.

MINIMUM MOVEMENT

Because a dividing head can rotate, the chances are that the work does not need moving by a large amount. In fact, virtually all dividing head work can be done with a movement of no more than 50 per cent of the maximum work diameter called for.

As mentioned earlier, to pitch out large diameters, where you cannot get the work under the dividing head, can be fairly easy. Just tilt the dividing head at 45 degrees and tilt the milling machine head by the same amount so they are facing each other. Use a test indicator to clock the large diameter so the dividing head is on the centre line of the spindle both horizontally and vertically, and all you need to do is move the milling machine by the radius of the PCD you require which will give you the PCD you desire. As the milling machine spindle remains on the vertical centreline it is a simple job to pitch out all the holes.

OTHER INDEXERS

Another simple type of dividing head is a spin indexer. The main type of spin indexer is a simple bracket with a spindle, a handle and a locking bolt. You simply put a pin through a division plate to index any amount up to 360 degrees by one degree increments. You can make a different index plate (simple to do from sheet metal) so that any number or position of holes can be indexed quickly and accurately.

The other simple dividing head is a bit more complicated; it allows you to index up to 36 positions by the simple operation of a lever. You can block off any of the 36 positions by tightening an Allen screw in any of the holes that do not require indexing.

Both of these indexers use C5 collets. The spin indexer can only be used horizontally (it can also be set in the vice or on an adjustable angle plate on small angles), but the other type of indexer can be used both vertically and horizontally.

Fig. 10.5. A spin indexer is ideal if the divisions required are full degrees. The division plate has holes around the outside while the pin can go in one of several holes to give a vernier type movement in whole degrees.

A spin indexer comes with a plate and holes in the body to divide a job into 1 degree increments. If only drilling some holes around the job, perhaps in an irregular pattern, you can make a replacement indexing plate, with only the required holes, to fit the spin indexer. I have done this on several occasions and it has saved loads of time and prevented scrap work.

SIMPLE DIVIDING BY 3, 4 OR 6

Using 5C collets has been mentioned earlier.

You can buy square and hexagon blocks to take 5C collets. You put the component in a collet in the block and then put the block into the vice to machine the work. You can work on any of the two, three, four or six faces by turning the block in the vice.

You can also make your own blocks that take the component direct by drilling a good fitting hole through the hexagon or square block, and drilling and tapping a grub screw hole to lock the component in place.

You can also put a bolt through a component and index from the hexagonal head of the bolt, or you can bolt a square plate to the component and index it by using a square to the flat face.

11 Drilling, Tapping and Reaming

DRILL CHUCKS

A drill chuck is essential on the small mill. It is ideal for holding drills, taps and reamers. However, it should never be used to hold milling cutters. The resulting side load or end load could cause the chuck to fly off its arbor, resulting in untold damage.

There are two main chuck types: the keyed chuck, which as its name suggests is opened and closed with a key, and the keyless chuck, which is opened and closed by hand.

The chuck will usually be fitted to a Morse taper arbor or possibly an R8 arbor. A useful alternative chuck mounting method is to fit the chuck onto a short parallel shank and hold it in one of the machine's collets, such as an ER25 collet. You could also use a side lock holder, which is especially useful if you have a small chuck for drills smaller than, say, 4mm. You can buy parallel arbors with a hand feed for these small chucks; the hand feed allows you to feed the drill into the work by holding a knurled ring on the adaptor.

Fig. 11.1. A keyless chuck can be tightened by hand.

Fig. 11.2. A keyed chuck, shown here on a lathe, uses a chuck key to tighten up.

CENTRE DRILLS

You are likely to come across many different types of drill in the home workshop. We will start with the two types of centre drill.

Slocombe centre drills

The Slocombe centre drill comes in various diameters, with the body usually starting at ⅛in diameter with a small pilot on the

Fig. 11.3. This keyed drill chuck has a shank that can be held in a collet. It is ideal for small drills up to 4mm. It can be fed into the work by hand feeding the knurled wheel into the job.

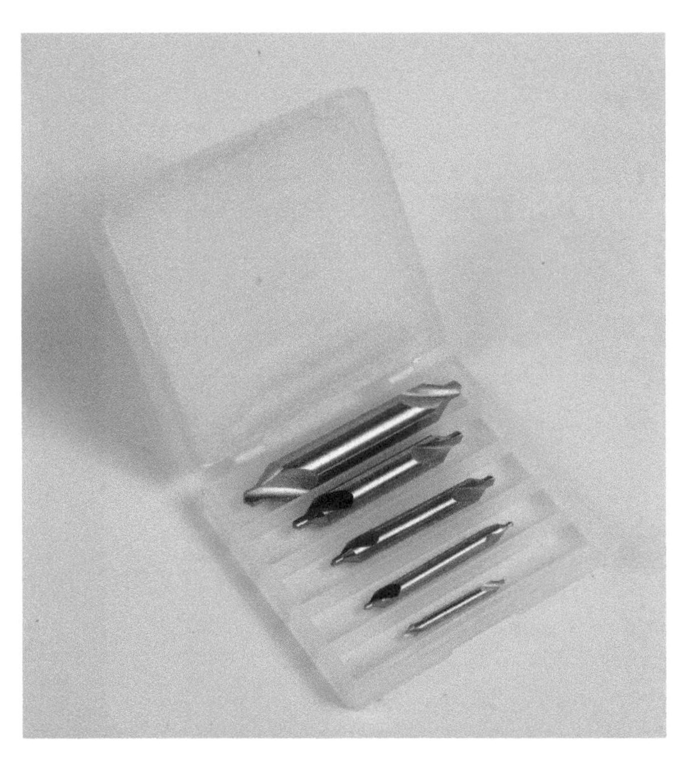

Fig. 11.4. A set of centre drills. Longer centre drills are available.

centre. The ⅛in size can be obtained with various different pilot diameters. You can also get centre drills in larger sizes such as ¼in, 5/16in, ⅜in and ½in. (Metric centre drills are also now available.)

Most centre drills are double ended. You can get long series centre drills, and I find if I cut these in half they are the ideal length for use in a chuck.

CNC centre drills

The second type of centre drill is called a CNC centre drill, so called because they are most often used on CNC mills. Rather than a pilot, these drills have a 90 degree inclusive angle so they can be used to centre to their full diameter. A CNC centre drill 10mm diameter can produce a centre spot up to 10mm diameter. This is very useful as, say, you want an 8.2mm diameter centre drill point to countersink a hole for an M8 screw, you just centre drill 4.1mm deep (half the centre diameter) and you get the correct size countersunk hole for an M8 thread. This diameter may vary slightly between makes of CNC centre drills, but 4.1mm would be a good start. Other sizes of countersink can be made by varying the depth of cut.

Fig. 11.5. A CNC centre drill will make a decent 90 degree countersink in the work.

Using an engraving cutter

Another type of cutter that can be used as a centre drill is an engraving cutter. You can get engraving cutters with various angles of point. These cutters are usually very sharp and if set right into the chuck or collet, they are very rigid. I find these engraving cutters are very useful as centre drills and they can be used to cut a centre on an angular face if used carefully. I have used this method on aluminium and brass components. I have not used them on steel but if care is used, they should be quite satisfactory.

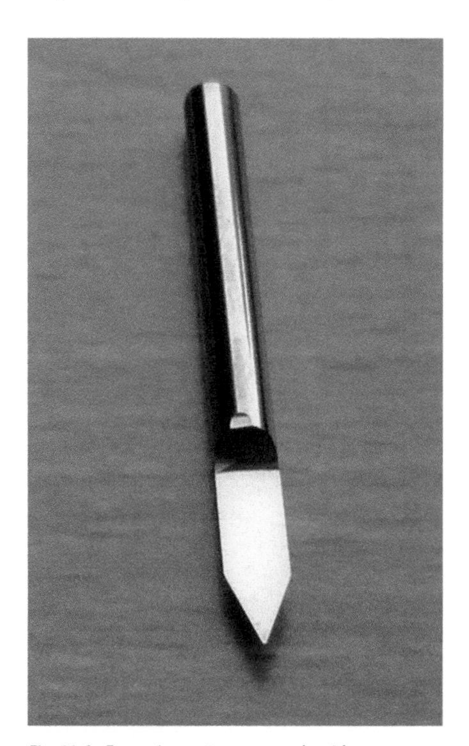

Fig. 11.6. Engraving cutters are made with many different cutting shapes. Something like this makes an excellent centre drill.

DIFFERENT TYPES OF DRILL

Twist drill

Until recently, the standard twist drill had not changed from when it was invented back in the early 1860s. The twist drill has spiral flutes and a standard inclusive angle of 118 degrees to the point.

Four-facet drill point

Recent advances in technology have seen the introduction of the four-facet drill point.

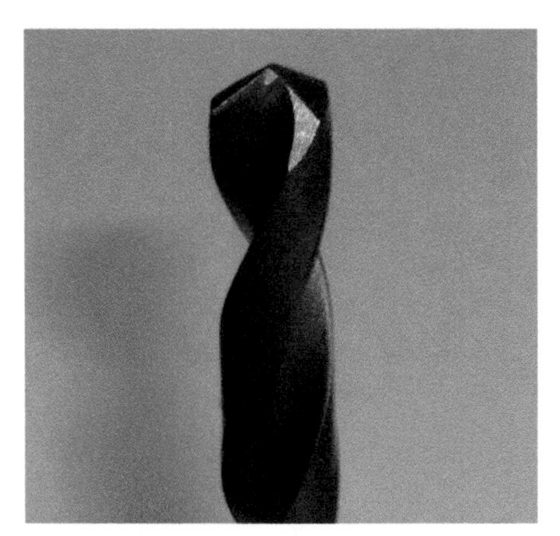

Fig. 11.7. A four-facet drill point.

Fig. 11.9. Stub, standard and long series drills. Longer drills are available.

This makes the drill self-centring, especially in the shorter or stub sizes. The four-facet point tends to cut better than a standard twist drill point. It is also far easier to sharpen by hand in the home workshop, although ideally a drill sharpening jig should be used.

Point angles

To drill plastics, the drill point angle could be increased to 90 degrees. If drilling harder metals, such as steel, which has possibly been heat treated, the drill can be obtained with point angles of 130, 135 and 140 degrees to give a longer life to the point.

Length of drill

Twist drills are available in many lengths from short stub drills up to very long drills for extra deep holes. If drilling a deep hole, it pays to start off with a short drill and increase the length of the drill as you drill deeper. This will enable the swarf to be easily removed.

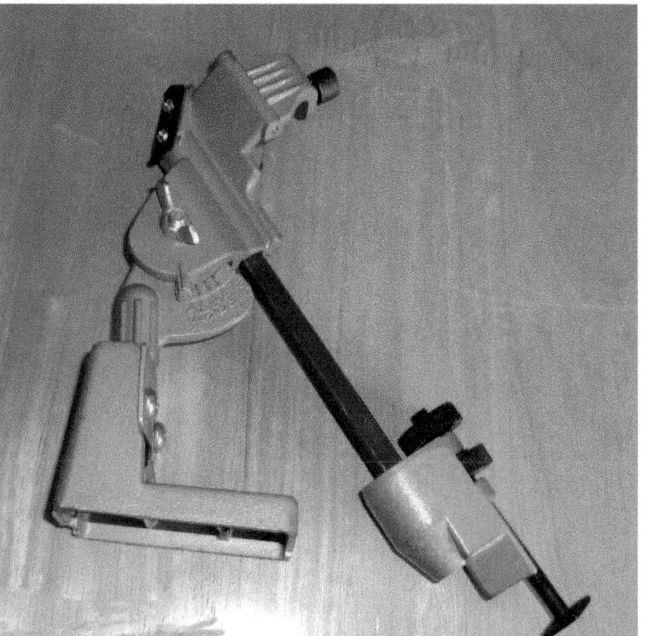

Fig. 11.8. A drill sharpening attachment for an ordinary bench grinder is widely available at low cost. After a bit of practice it will be easier to sharpen drills with this device than trying to do it freehand.

Fig. 11.10. Core drills are used for opening out holes in castings. They usually have three or more flutes.

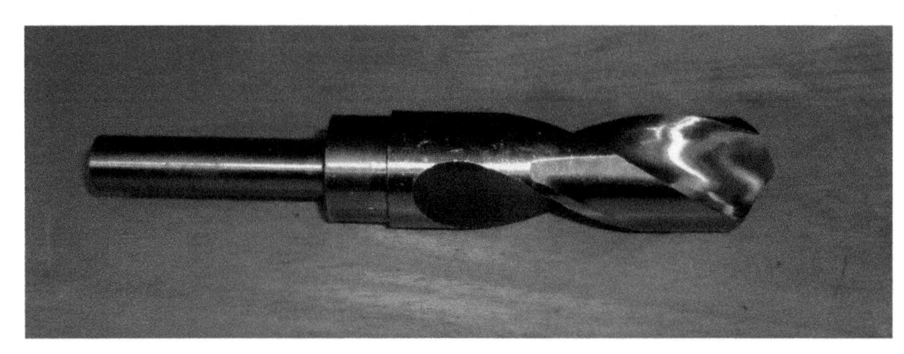

Fig. 11.11. Blacksmith's drills normally have a ½in shank and can be held in a chuck or collet.

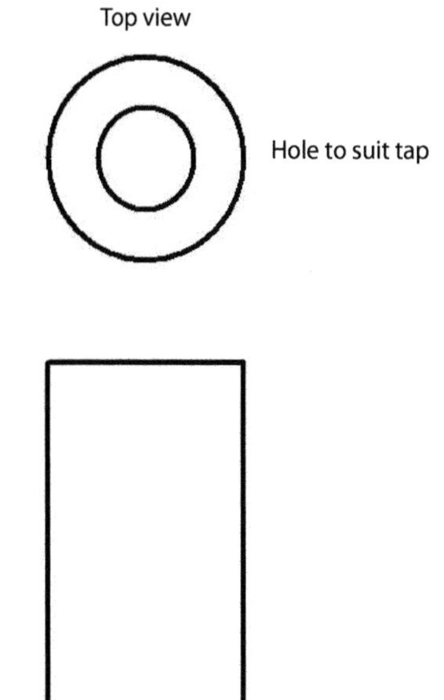

Fig. 11.12 A tapping guide can be made from a bit of round bar. Drill and face it at the same time and put a small groove near the end that you faced so you know which way up to put it.

Extensions

It is possible to extend a standard drill by fitting an extension to the drill's shank. Just drill a hole to suit the drill shank into a round bar and fix the drill into the hole using Loctite (an engineering adhesive) or silver or soft solder.

Blacksmith's drills

Blacksmith's drills usually have a large head and a small (usually ½in) shank. They are available in many sizes.

TAPPING

Tap centre support

Often, when you have drilled a hole it needs to be threaded with a tap. To keep the tap square to the hole, you need to support the shank of the tap in some way.

Tap shanks are usually ground to a point or have a centre in them. To support the end of the tap, turn a 60 degree inclusive centre onto a bit of round bar and skim the diameter of the bar to fit a standard collet. This will ensure concentricity. Do the same with another bit of bar, but this time centre drill the end of the bar leaving a 60 degree cone to accept the end of the tap.

An alternative method to hold the tap vertically is to drill a bit of bar to take the threaded bit of the tap and use this on the work to support the tap upright.

Tap wrenches

I suggest you get a few different tap wrenches – a couple of the tee type wrenches and a couple of standard tap wrenches to suit the size of work you will be doing.

Fig. 11.13. A selection of tee type and standard tap wrenches.

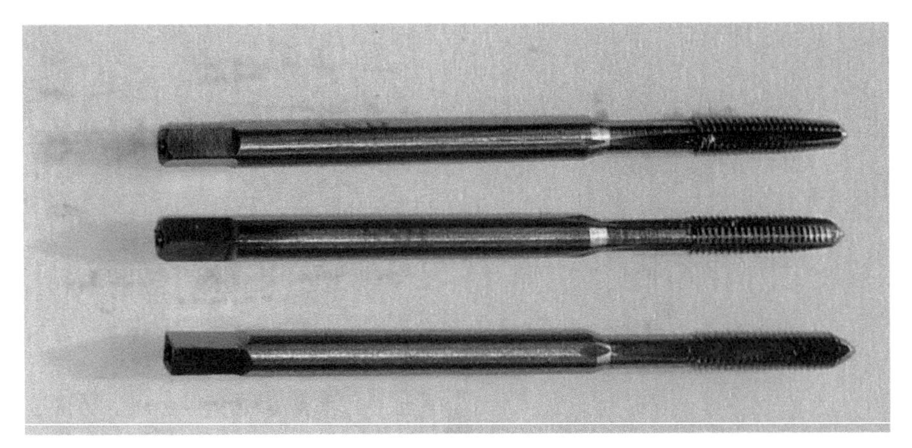

Fig. 11.14. A set of hand taps.

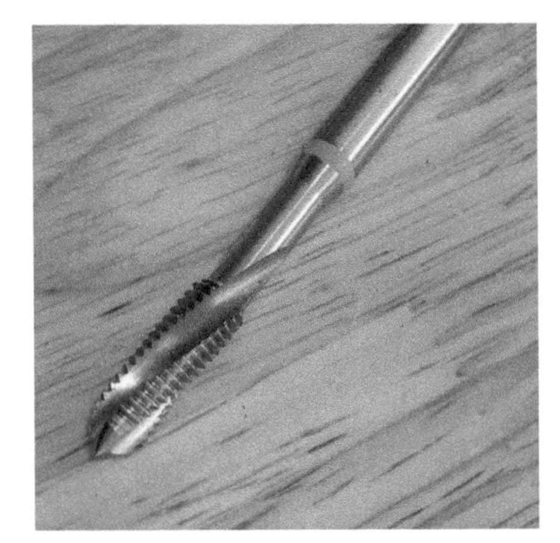

Fig. 11.15. A machine tap.

Types of tap

For our purposes, you will probably use either hand taps or machine taps. The hand taps will be used with a tap wrench as mentioned previously, while machine taps will probably be put through the work under power but only if the tapping hole is drilled right through the work.

REAMERS

Types of reamer

As with taps, you will probably come across hand and machine reamers. Hand reamers can be used the same way as taps with a shank support. A hand reamer normally has a tapered lead on its front to lead the reamer into the hole.

A machine reamer has a parallel cutting diameter and needs to be held in the machine's chuck or collet or possibly in the machine's Morse taper, depending on the type of shank it has.

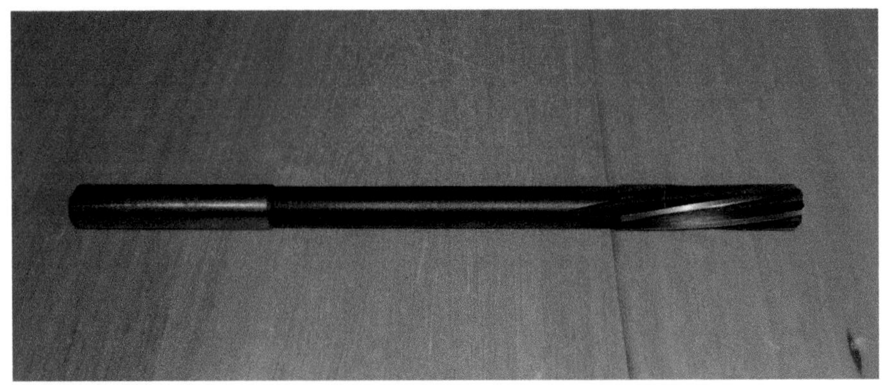

Fig. 11.16. A machine reamer has a parallel cutting edge.

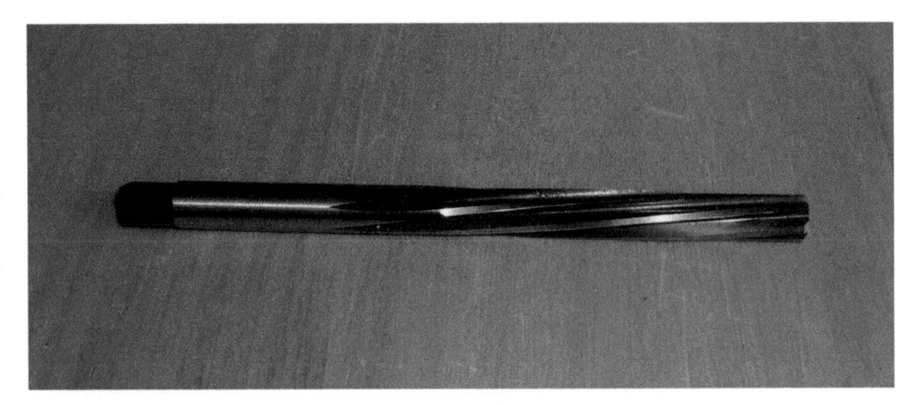

Fig. 11.17. A hand reamer has a tapered lead to the cutting edges.

Using a reamer

When preparing a hole for reaming you should use the two-hole drilling method.

You need to drill a decent size hole prior to reaming. For small reamers up to ⅜in (10mm) leave about 0.012in (0.3mm) for the reamer to remove. For reamers up to ¾in (19mm) leave 0.016in (0.4mm) and over ¾in (19mm) leave 0.020in (0.5mm).

For reaming speeds use about ⅓ to ½ of the drilling speed for the same diameter. Feeds should be from 0.002 to 0.004in (0.05 to 0.1mm) per tooth per rev. Multiply the feed per tooth by the number of teeth to give the feed per revolution. When you are feeding the reamer you will probably feel it if you are pushing the reamer too fast as it will start to bind up.

Use plenty of neat cutting oil for steel, brass, bronze, gunmetal and cast iron, but for aluminium paraffin should be used. For any type of plastic you should not use oil as a cutting lubricant as it can attack and/or degrade the plastic; instead compressed air or plain water should be used.

DRILLING ADVICE

Drilling 90-degree cross holes

You can drill 90-degree cross holes through a bar by using a bar through the first hole resting on two parallels or clocked true with a test indicator. You could also drill cross holes using a dividing head.

Fig. 11.18. Exaggerated view of using an offset cutter to enlarge a hole. The blue is the original hole and the red shows the four offset holes. If you look carefully, you can see the lobbing at the top and bottom and left and right.

Enlarging a hole by offsetting the cutter

Sometimes you will not have a drill large enough to drill a hole of the size required. In this case, it is possible to open the hole out larger with a slot drill or end mill. Say the hole required is 22mm in diameter and the largest cutter you have is 20mm; all you have to do after drilling the hole is to put the cutter through the hole at the centre point then put it through the hole at 0, 90, 180 and 270 degrees, all offset by 1mm. Then you can finish the hole by boring. This is a lot quicker than trying to remove 2mm from the hole with a boring bar.

PITCHING OUT HOLES USING CO-ORDINATE DRILLING

Set up your component on the milling machine's table and select a datum for the machining. If you have a digital readout, all you have to do is carefully pitch out the holes and they should be accurate. However, if you do not have a digital readout the accuracy may be suspect unless you move from co-ordinate to co-ordinate in such a way that avoids backlash becoming a problem.

The illustration shows all the required co-ordinates for drilling the hole positions for a small steam engine bedplate casting. The machining datum is the top left intersection of the two datum lines. This is where all the co-ordinate dimensions are taken from. I have also added letters to the hole centres to show the drilling sequence I used.

Begin with the wobbler

First you have to find the edge of the back face and then the left-hand end of the casting in turn. Wobble off the back face first, moving the table very slowly until the wobbler rolls along the edge of the casting. Raise the spindle and move the table half the diameter of the edge finder. Zero the dial (or make a note of the reading of the dial if you cannot zero it). The centre of the edge finder should be over the edge of the job.

Move the table another turn so that the edge finder is over the job; the distance is unimportant as long as the edge finder is on the end of the casting and not the corner radius. Now you need to use the edge finder to find the left-hand edge, then move the table half the wobbler diameter towards the middle of the casting. Zero the dial for the axis for the edge you have just found. Wind back in the first axis you set so the edge finder is a turn or two away from the job. Wind back to zero so the centre of the spindle is over the datum point. You are now at the theoretical intersection point of the top left corner of the job.

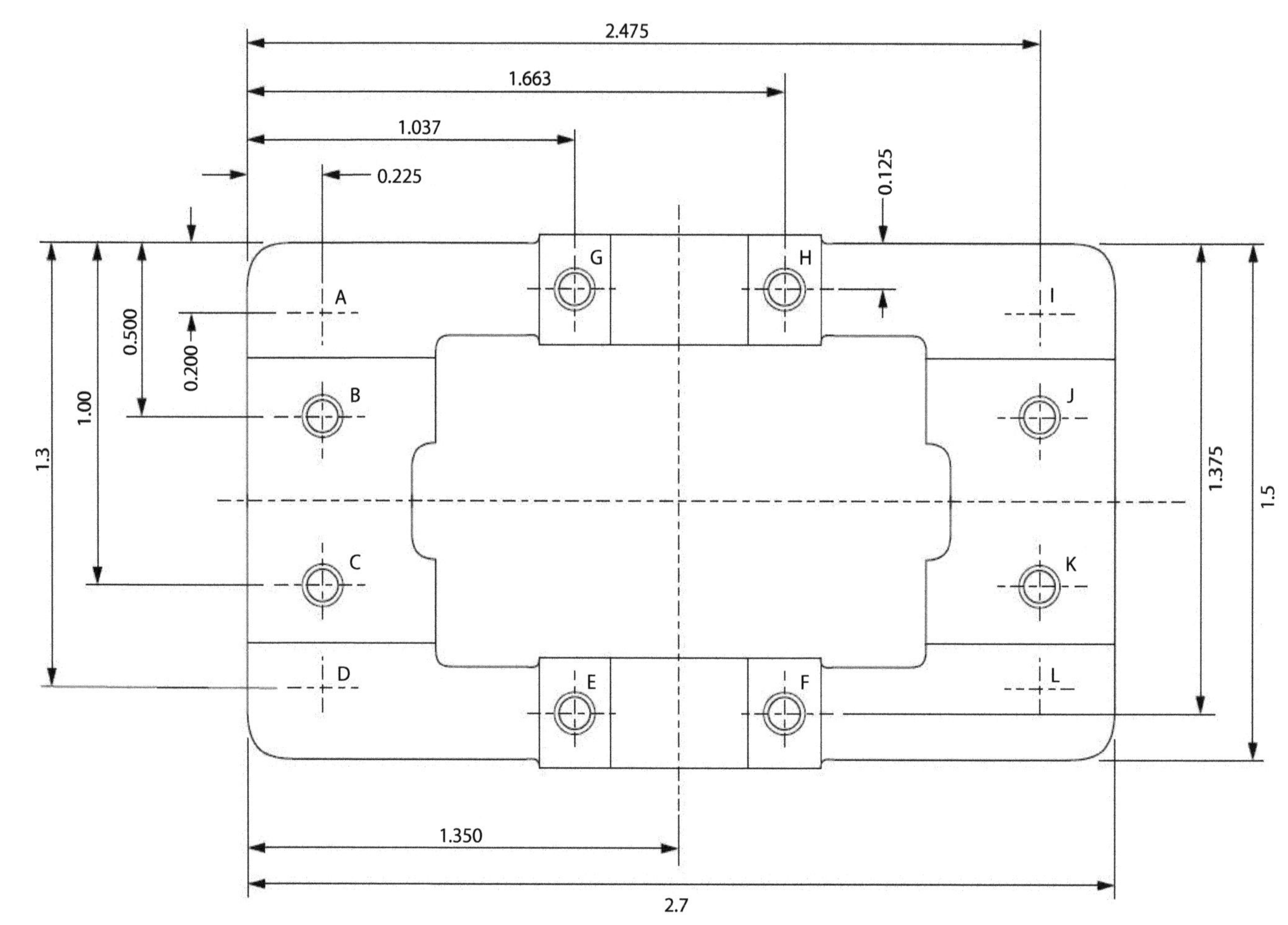

Fig. 11.19 Make a drawing before pitching out holes. It does not need to be high quality, just an accurate sketch will do.

Centre drilling holes A–F

Now we can start to pitch out the holes. Move the table in the longitudinal axis to 0.225in so you are in line with the row of four holes at the left of the job. Centre drill the first hole. This hole may wander slightly as it has already got a dimple from the casting process. You should not be too far away from the centre of the dimple though. Move in the cross axis to 0.5in and centre drill the next hole. Move to the 1.0in position and again centre drill the hole. Move to the 1.3in position and centre the last of this row of holes.

Move to the 1.375in position and then move to the 1.037in axis in the longitudinal direction and centre drill the hole. Finally move to the 1.633in position and drill this hole. We now have holes A, B, C, D, E and F centre drilled.

Fig. 11.20. Spot the first of the holes.

Centre drilling G–L

Now we go back to the start point at the top left of the component. Move over to the 0.125in position and move along to the 1.037in position. Centre drill the G hole, then move along to the 1.633in (H) hole and centre drill again. Now you can move over to 0.2in in the cross axis and to 2.475in in the longitudinal axis. Centre drill the I hole and move over in the cross axis to centre drill holes J, K and L.

Fig. 11.21. Spot the rest of the holes.

Eliminating backlash

You have now drilled all 12 holes in the correct position without any errors from backlash in the feed screws. You can always eliminate backlash by going back a couple of turns and then moving to the starting point again. As long as you only drill holes when you are moving the same way, the holes should always be in the correct position. If you did accidentally wind past the required hole position, just wind back a turn or two and then move up to the hole again. By going back a couple of turns, and then moving forward again, you are taking the backlash out of the equation and ensuring you are drilling the holes accurately.

Review the method

If this is not all clear, read it through again doing the moves in your mind using the drawing. The principle will work for any component you care to machine. Yes, you might have to redraw the component out differently or redo the dimensions from the datum rather than from an existing datum or centre point, but it will save you time and stop you from scrapping work. Do make sure you double check your dimensions carefully if redrawing the component's co-ordinates.

As well as being relevant to machines without a readout, I have used the same principle of reducing backlash on a worn-out CNC machine and reduced errors to within drawing limits.

12 Boring Holes and Repetition Work

BORING TOOLS

Boring tools are used to clean up a previously drilled hole, possibly to make it run true but more likely to ensure the hole is to the correct size. Boring tools are normally longer than ordinary tools and tend to stick out further from the tool holder. HSS tools are available and are suitable for most uses.

Different styles of boring bars

The two most useful types of boring bars will be the high-speed steel ones (HSS) and the carbide-tipped ones, either brazed-on tips or inserted tips. Carbide inserted tip-boring bars are the best bet for boring cast iron in the small workshop.

You can also get boring bars with mild steel shanks and inserted high speed steel tool bits, usually in the form of short round bars held captive with a grub screw.

For the really small holes, solid HSS and carbide boring bars are readily available. Although they are not cheap, they will be found very useful and quick to use. They go down very small – holes and 2 or 3mm diameter bores are possible. I often buy these little tools in second-hand tool shops or on eBay for a fraction of their new price.

Another method of boring a hole is to use a long series slot drill as a boring tool. In use, clamp the shank in the boring head, making sure the rest of the slot drill is clear of the side of the bore. You can then run it through the bore similar to a standard boring bar.

If you just need to true up a hole, where the size is not important, you can just run the slot drill straight through the hole while holding it in a collet. If you have a hole that needs reaming, a boring bar or slot drill will be the ideal method of bringing the hole nearly to size while also truing it up. An old slot drill could be backed off and the remaining tip stoned or lapped down slightly undersize on the nominal diameter, leaving just a few thou for finish reaming.

Fig. 12.1. Small high speed steel boring tool.

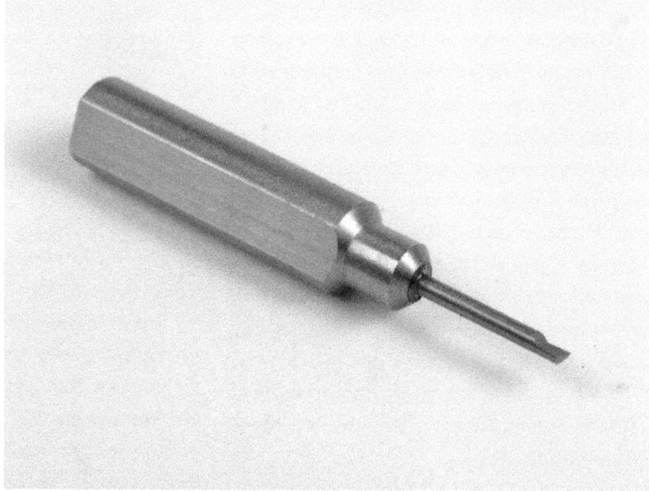

Fig. 12.2. Small brazed carbide boring tool.

Fig. 12.3. A typical boring head with a carbide-tipped boring bar.

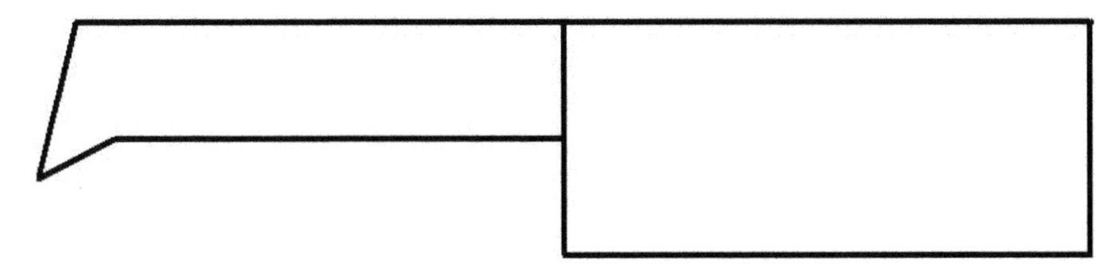

Fig. 12.5 Boring bar with trailing edge to suit blind holes.

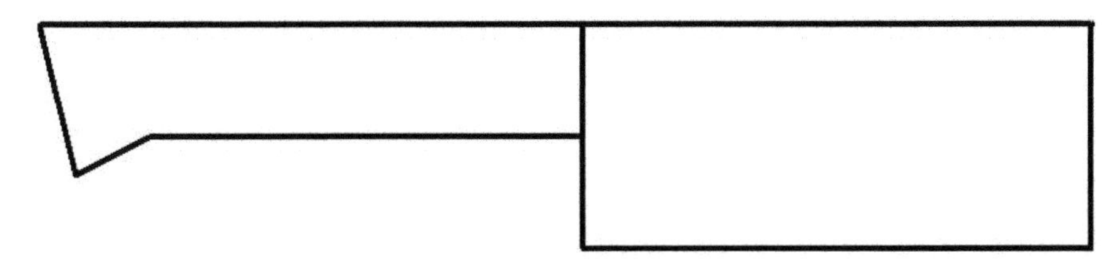

Fig. 12.6 Boring bar with leading edge suitable for through holes.

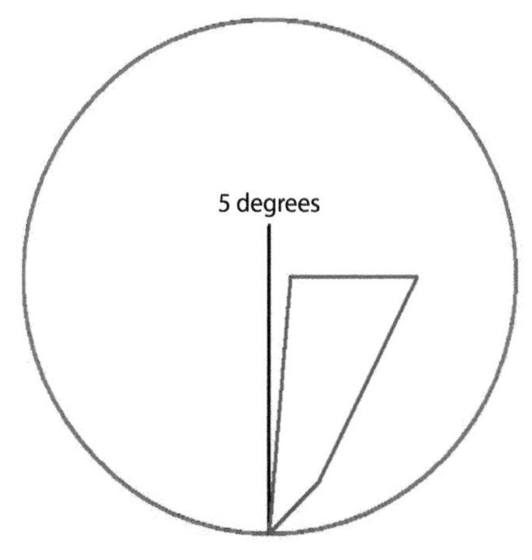

Fig. 12.4 Correct orientation of cutting edge in relation to the hole. The tool has a 5 degree rake angle on the front face of the tool. This angle should be fine for most jobs.

Boring heads

A boring head usually has two or three holes to take the boring bars. Some boring heads also have a hole in the side for use when boring large holes.

The length of the boring bar required will depend on the depth of the hole to be bored. Keep the boring bar as short as possible to maintain the stiffness of the tool.

In use, the boring bar should be set the same as a lathe tool so the cutting edge is presented to the hole being bored at the correct angle. A boring bar is much more liable to chatter in a boring head than in a lathe, so it is essential that the tool is very sharp.

If the hole being bored is a through hole the boring bar can have a forward leading edge, but if the hole is a blind one it needs a trailing edge so the end of the boring bar does not hit the bottom of the hole.

Checking a bore for accuracy

A bore usually needs to be an exact size so we need a method of measuring the bore. The simple way could be to use the actual component that fits into the hole if it is short enough, but this may not be possible

Fig. 12.7. A dial indicator gauge suitable for checking a bored hole.

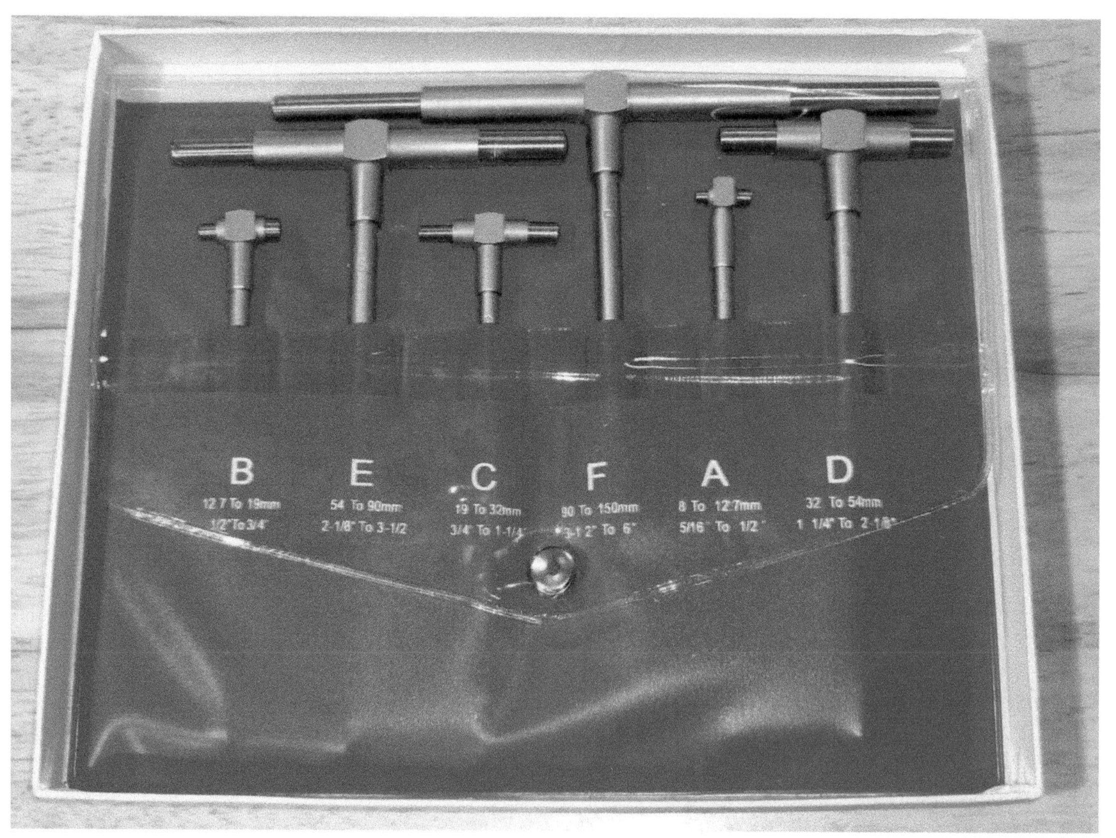

Fig. 12.8. A set of telescopic gauges suitable for checking bored holes.

if the component is very long and difficult to handle, in which case we have to make a gauge or use a measuring instrument.

If we cannot use the component or a gauge, a good alternative is a telescopic gauge. These are simple to use and very accurate. You put the telescopic gauge into the bore on an angle and lightly nip the locking screw. Then carefully straighten up the gauge in the hole. The gauge will close down to the exact diameter of the hole, and you can then tighten it up and measure it with a micrometer. This will give you an accurate reading of the hole size. It is advisable to run the boring bar through the hole a couple of times at the same setting to ensure any deflection is removed.

Boring a seating for a chimney

Cutting a scallop, for instance to shape the bottom of a chimney to fit the boiler barrel, is an easy job with a boring head. Set up the chimney horizontally with the scallop end nearest to the boring bar. Set the boring head centrally to the centre line of the chimney.

To set the correct radius on the boring bar find the edge of the bar's face. Move the table away by the 100mm boiler barrel radius so you are exactly 100mm away from the workpiece's face.

Put the boring bar and tool into the machine's spindle and open out the boring head until the boring tool just touches the bottom face of the chimney. The boring bar is now set to the exact radius of the boiler barrel, 100mm.

To cut the radius, just move the machine's table a few thou towards the chimney and take a downwards cut to clean up the bottom of the chimney. Continue to do this until the radius on the bottom of the chimney is fully formed. Note: Do not alter the radius of the boring bar once it has been set.

You should now have a chimney that is a perfect fit on the boiler barrel. This method can be used for all sorts of fittings that need to be fitted to a specific radius.

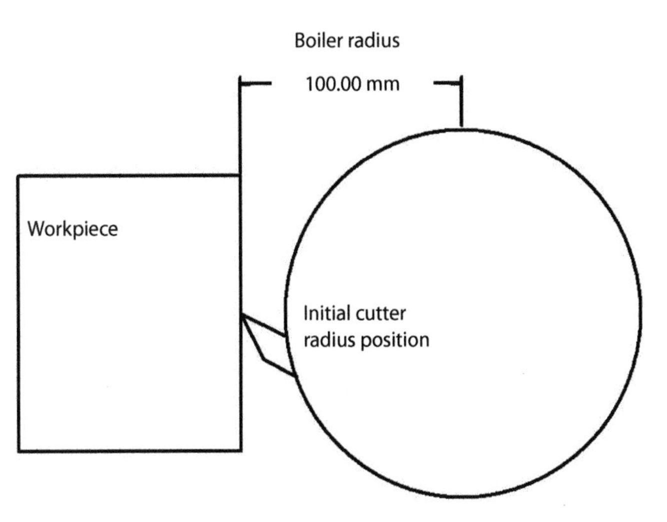

Fig. 12.9 The initial setting of the chimney.

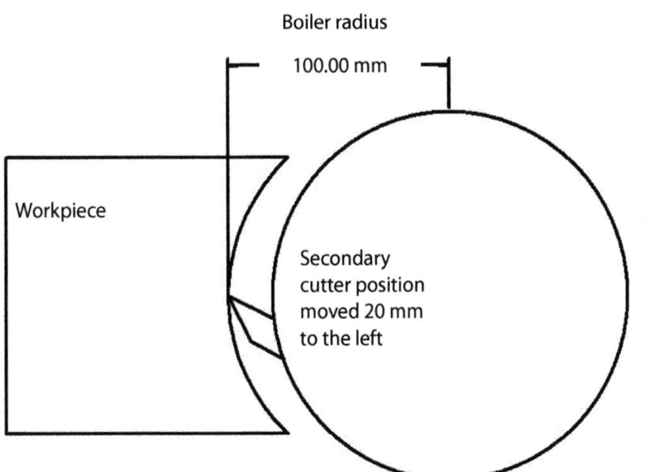

Fig. 12.10 The machined scallop on a chimney base.

Trepaning

If you need to bore a hole in a relatively thin sheet of metal you can trepan the centre out. Trepaning is a method of cutting a groove through a piece of metal so the middle can be removed. The tool should cut on the bottom, and both sides of the tool should be relieved. Feed the tool slowly through the metal taking particular care when the centre breaks through. The hole can then be finished using a conventional boring tool.

Boring outside diameters (turning)

You can bore outside diameters in the mill. This is the equivalent of turning. You do this by putting a suitable tool in the boring head. If you have a left-hand boring tool, use this, but if not, you can use a normal boring tool and run the machine in reverse. This method is ideal if you want to machine a spigot on an odd-shaped component and do not want the bother of setting it up in the four-jaw chuck on the lathe. Start with the boring bar clearing the outside diameter of the material and gradually reduce the boring head until the diameter required is met.

REPETITION WORK

Machine stops

Machine stops are very useful especially if you do not have a digital readout. Not all machines have table stops as standard, but they are not that difficult to fit either. The mill's spindle will almost certainly have a depth stop to control the depth of drilling.

If the milling machine has a knee that can be wound up and down, this is the quick way to set the depth stop. Simply put the drill into the chuck or collet and wind the spindle down to the stop; then wind the table up until the drill touches the top of the work. Now we know where we are, we can raise the table by the required drilling depth, and the point of the drill will drill to the correct depth.

If you do not have a knee, you can put a drill shank or block of metal on the drill stop and lower the head or otherwise adjust the depth of the drill point until the block is captive and the drill point is on the top of the work. Remove the block and you can drill to the required depth.

Traverse stops

Traverse stops on the X and Y axes are very useful. It is easy to wind past the required position, but if you have a stop in the way you cannot. Set the stops at the extremes of the travel by using a block between the stops or by measuring the gap with a vernier calliper.

DTI on stops

A plunger dial test indicator can also be fitted to the stop blocks so you can see when you reach zero. These plunger indicators can usually have the dial set to zero so you can make fine adjustments to get the position accurate.

Fig. 12.11. Stops on the X and Y axes will be found very useful for repetition work. The two stops on the right-hand side can be moved anywhere on the front of the table. The bar to the left is the stop bar, and on this machine the stop bar can be raised or lowered.

Fig. 12.12. Simple milling fixtures can often save time in the home workshop. They do not need to be complex, simple is best. This fixture is a simple block with a turned register to take the cylinder castings. The bush is to clamp the cylinder down.

Milling fixtures

Simple milling fixtures can be made from a piece of plate, a few positional blocks and a couple of clamps. If doing more than one component the fixture will often save you a lot of time. You can keep changing the components and do each operation one at a time until all the machining is completed.

Fig. 12.13. The cylinder fits over the spigot. Note the small hole in the block: this will take a location pin when drilling the second end.

Fig. 12.14. After drilling the first set of holes the casting is removed, a small pin is put into the fixture and the cylinder put back on the other way up to drill the holes in the other end. The location pin ensures the holes are in line at both ends.

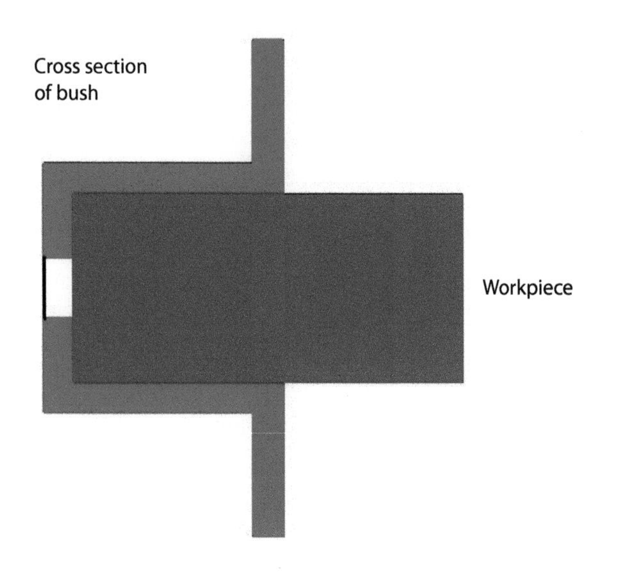

Fig. 12.15 A cross-section of a top hat bush being used as a backstop in a chuck. The collar goes against the face of the chuck. This bush could be used vertically or horizontally.

Back stop in collets and chucks

Fit an end stop into the end of a collet or chuck to ensure every component locates in the same way. If you cannot use a back-stop, use a top hat bush where the rim of the top hat goes against the face of the collet or chuck.

AND FINALLY...

We have come to the end of the book. Hopefully you are wiser and you can tackle any milling job you might encounter in your home (or place of work) workshop.

Most engineering is a matter of common sense, care and patience. Before you start a component, think about it, make a sketch and write down the sequence to make it. Ensure you have all the tools required to hand.

If the component is a funny shape or you do not understand the drawing, make a mock-up by hand with plasticine. It may seem laughable, but this method works. I have modelled sewing machine components and aircraft parts to figure out what I am trying to make. Plasticine is cheaper than metal.

Enjoy your milling machine and make sure you use safe working practices. Remember, milling cutters are sharp!

Glossary

bezel – the outer rim of a dial

CAD – computer-aided design

CNC – computer numerical control

collet – a device for holding components

cutter – a tool for removing or slicing

datum – the initial zero point

HSS – a wear-resistant hardened steel

involute – curved spirally

mandrel – a device to mount work on

quill – the movable part of a milling machine's head

ram – the centre moving part of a turret mill

reaming – enlarging a hole using a ream

shim – a thin component to make up the size of another component

swarf – the metal waste removed by a cutting tool

thou – one thousandth of an inch

tram/tramming – truing up the milling machine's head

vernier scale – a scale for dividing the main scale into smaller divisions

Further Information

USEFUL CONTACTS

Model Engineer and *Model Engineers' Workshop* magazines, www.model-engineer.co.uk

SUPPLIERS

UK

Allendale Electronics Ltd, Machine DRO Dept, Pindar Road, Hoddesdon, Hertfordshire, EN11 0BZ, Tel: +44 (0)1992 455921, www.machine-dro.co.uk
Suppliers of digital readouts and measuring equipment.

Arc Euro Trade Ltd, 10 Archdale Street, Syston, Leicester, LE7 1NA. Tel: 0116 269 5693, Email: information@arceurotrade.co.uk, www. arceurotrade.co.uk
Suppliers of lathes, manual and CNC milling machines, tools and accessories.

Blackgates Engineering, Unit 1, Victory Court, Flagship Square, Shawcross Business Park, Dewsbury, WF12 7TH. Tel: 01924 466000.

Chester Machine Tools, Hawarden Industrial Park, Hawarden, Chester, CH5 3PZ. Tel: 01244 531631, Email: sales@chestermachinetools.com, Website: www.chestermachinetools.com
Suppliers of machine tools and accessories.

Cowells Small Machine Tools Ltd, Tendring Road, Little Bentley, Colchester, Essex, CO7 85H. Tel: 01206 251 792, Email: sales@cowells.com, Website: www.cowells.com
Suppliers of the Cowells range of lathes and milling machines including a clockmaker's lathe.

Model Engineering Services, www.modelengineeringservices.com
Suppliers of castings for the Quorn and Kennet tool and cutter grinders.

Transwave converters at Power Capacitors Ltd, 30 Redfern Road, Tyseley, Birmingham, B11 2BH. Tel: 0800 035 2027, Email: transwave@powercapacitors.co.uk, Website: www.transwaveconverters.co.uk
Inverter suppliers and converter manufacturers.

Warco, Warco House, Fisher Lane, Chiddingfold, Surrey, GU8 4TD. Tel: 01428 682929, Email: sales@warco.co.uk, Website: www.warco.co.uk
Suppliers of machine tools and accessories.

USA

Grizzly Industrial Inc., Bellingham Washington, 1821 Valencia St., Bellingham, WA 98229. Tel: 1-800-523-4777, Website: www.grizzly.com
Suppliers of machine tools and accessories.

Harbor Freight, 3491 Mission Oaks Blvd, P.O. Box 6010, Camarillo, CA 93011-6010. Tel: 1-805-388-3000, Website: www.harborfreight.com
Suppliers of machine tools and accessories.

Australia

Carba-Tec, 151 Balcatta Road, Balcatta, WA 6021. Tel: 08 9345 4522, Email: perth@carbatec.com.au, Website: www.carbatec.com.au
Suppliers of machine tools and accessories.

Hare & Forbes Machinery House, The Junction, Unit 1, 2 Windsor Rd, Northmead NSW 2152, PO BOX 3844, Parramatta NSW 2124. Tel: (02) 9890 9111, Website: www.machineryhouse.com.au/Stores
Suppliers of machine tools and accessories.

New Zealand

Carba-Tec, 110 Harris Road, East Tamaki, Auckland. Tel: 09 274 9454.

Index

RELATED TITLES AVAILABLE FROM CROWOOD

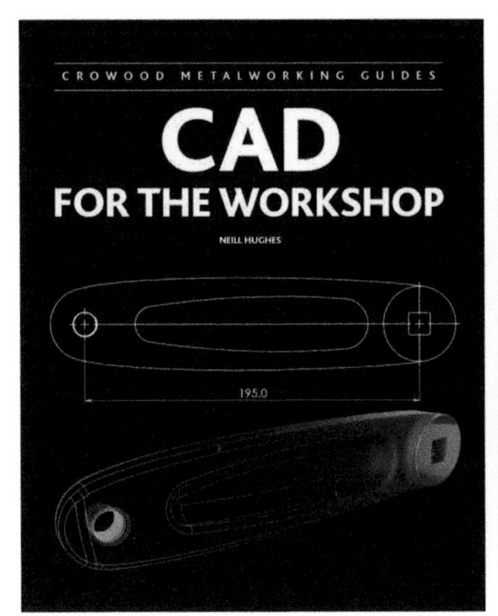

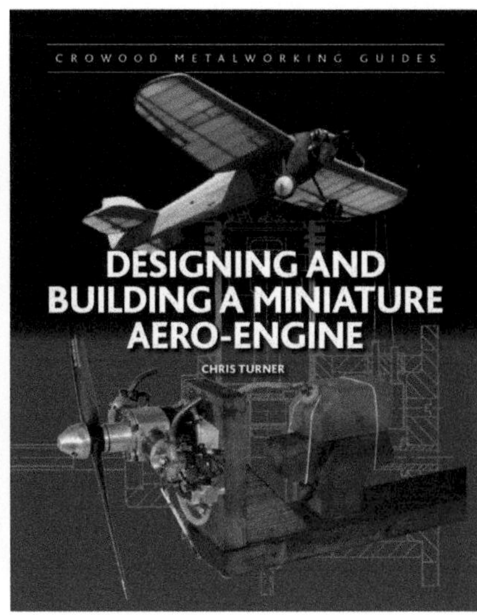

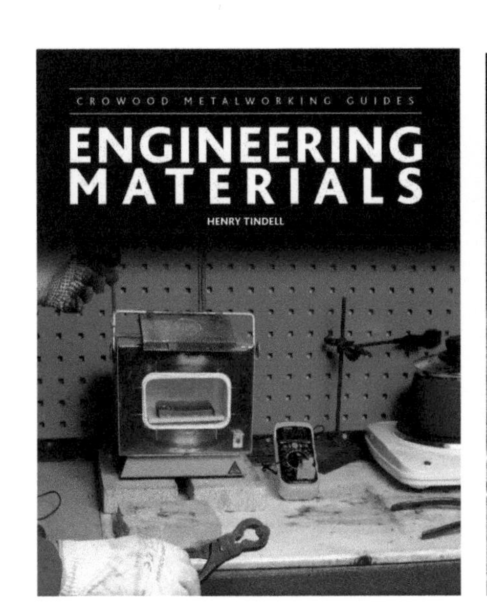

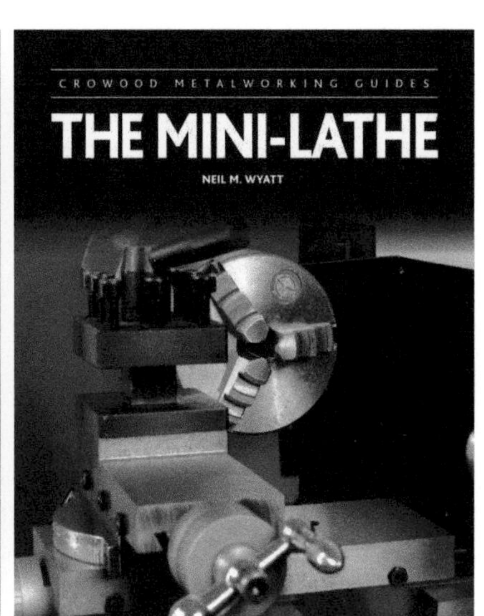